SOLVED

A TEACHER'S GUIDE TO MAKING WORD PROBLEMS COMPREHENSIBLE

SOLVED

A TEACHER'S GUIDE TO MAKING WORD PROBLEMS COMPREHENSIBLE

DIANE KUE

atmosphere press

Published by Atmosphere Press

Cover design by Beste Miray Doğan

atmospherepress.com

CONTENTS

CHAPTER 1

WORD PROBLEM PERSPECTIVES

While I was a classroom teacher, I was introduced to so many quick-fix strategies for students who struggled with word problems. One year my campus pushed for all students, whether they struggled or not, to use the same word problem-solving strategies. The justification was that we were teaching them life-long skills they would use when they were older and problems got harder. They included but were not limited to: read the problem at least twice, circle key words, underline important numbers, and rewrite the question. My students hated word problems and even without research, I understood why. Students who understood how to solve them loathed the extra, tedious work. The students who struggled with word problems were not receiving support that met their specific needs. Instead, I was giving them a one-size-fits-all method that did not help them build conceptually. I was not teaching the student; I was teaching based off an understanding that every student had the same needs on how to solve a word problem.

Because students are diverse, it is difficult to meet all students' needs during instruction. Thus, the pedagogy behind teaching word problem comprehension is commonly replaced by creating and implementing quick-fix strategies addressing misconceptions. Here are some misconceptions on how to comprehend word problems, and what the research says:

Misconception	Research says
Teach students key words so they know what operation to use. Example: *times* means *multiply*	Using keywords as a replacement for comprehending the problem and number relationship subverts the problem-solving process and is not sustainable or foolproof. Additionally, when problems become complex and involve multiple relationships between quantities, searching for keywords is not reliable (Clement and Bernhard 2005).
Students who read below grade level do not have the skill to comprehend word problems and need to read the problem multiple times to understand.	When learners collaborate with a partner, small group, or large group, the culmination of varying ideas and perspectives becomes a resource for the text (Beers and Probst 2017). The learners become responsive readers who use socialized learning to scaffold comprehension as they interact with the text and find meaning.
Students should solve the problems by themselves because they are not allowed help on tests.	When learners are expected to describe their strategic process to their teacher or within a group, they obtain higher mathematical achievement (Webb et al. 2014).

The misconceptions contrast with research because they do not have the same entry-level perspective. For example, two teachers with differing perspectives or understanding upon entering a professional learning community—one with the notion that students must always work alone to solve word problems and another who believes students should work together—will approach teaching word problems differently because their entry-level perspectives are different. Finding that common perspective from which to build upon is critical to understanding one another and building on concepts. Instead of applying strategies, educators must be *strategic* in their approach to teaching. This is where the transformation or the shift in the mathematical pedagogical approach begins.

Reflection Checkpoint
What approaches help you teach how to solve word problems? What doesn't help?

CALCULATIONAL VS. CONCEPTUAL ORIENTATION

The pedagogical perspective to teaching word problem comprehension comes in two forms: calculational orientation and conceptual orientation (Thompson et al. 1994). In a nutshell, calculational orientation means you may not know why the math is the way it is, but you can do it. For example, to find the hypotenuse of a right triangle, the formula is $a^2 + b^2 = c^2$. However, do you know how this formula works, or why it is true? Conceptual orientation means you understand why and how the math works and is applied. To discover the formula for finding the hypotenuse of a right triangle, a teacher might have students experiment with a leading question such as, what is the relationship between these squares and this triangle?

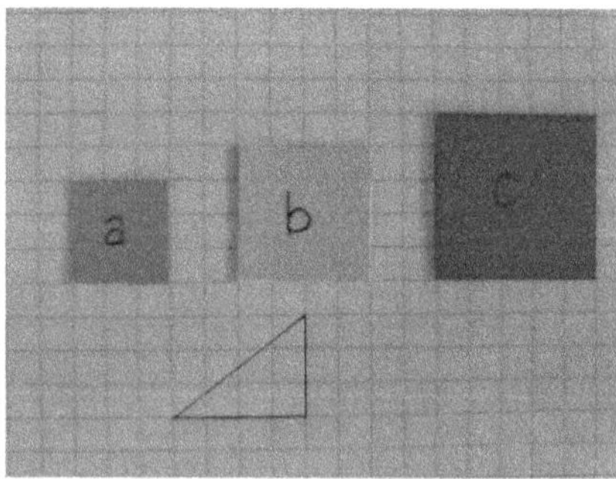

As students manipulate the squares, they might discover something you have never noticed, and they will also most likely discover this:

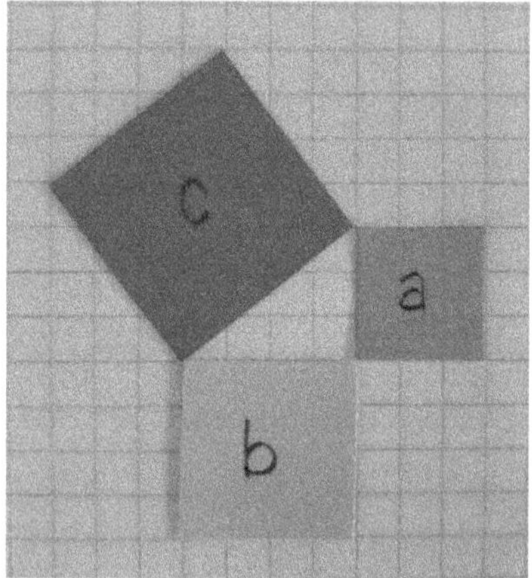

From that discovery, the teacher can guide students as they build on this idea. Can it be repeated with different sized squares? What is the relationship between the values of the squares? Does this relationship apply to other types of triangles? How can this relationship be expressed?

In a classroom setting with a limited mathematics block, it might seem easier to take the calculational route. When I taught using calculational orientation, my lesson was very much like this: Here is the content objective. We will apply it to these twenty problems that you will do in class, and what you don't finish will become homework. I will meet with a small group for intervention while the rest of you work on these problems individually.

In a conceptual orientation lesson, a whole mathematics block might be used to build the concept without students solving any additional problems. For the teacher, it might appear or feel like you are behind. I hope that even with the pressure of teaching and getting so much mathematics done in one class period, you still consider the conceptual approach for the sake of how much it benefits your students' foundational knowledge. I spoke with my seventeen-year-old son, who at the time of our conversation was a junior in high school taking a college statistics course. On the topic of conceptual learning, he said, "My world feels so much more open when teachers allow me to do things, like being told a question and told to figure it out. Quite plainly, I had no clue what a variable was until my ninth-grade teacher, who when she taught variables asked us what a variable is. Before that moment, I always just solved for an equation. I never thought about what it meant. When a teacher skips steps, I skip those same steps. I have no idea what I missed because I was never taught it."

Using conceptual orientation, students discover and find those insightful steps that furnish them with life-long skills. Educators encourage attainment of mathematical understanding. On the next page is an example of how the two orientations manifest within the same word problem.

I learned from my uncle that tall people tire of talking about their height. When strangers ask him his height, he always responds, "I am 1 yard, 2 feet, and 20 inches." How tall is my uncle in feet and inches?

Calculational	Explanation
What is the question asking? What are the numbers and units of measurement in the problem? How can you calculate his height?	Questioning is exclusively about the numbers, units of measurement, and what operations are needed to solve the problem.
What do you need to do first to calculate his height? What do you need to do second?	The emphasis is placed on mathematical procedure to find an answer.

Conceptual	Explanation
How does the uncle choose to reveal his height?	Questioning promotes ideas and development of ideas.
What relationship do these units of measurement share? How does understanding this relationship help determine the uncle's height?	Focus shifts from procedure to exploring concepts, ideas, applicable situations, and relationships.

Use your conversion chart.	There is disregarding of context, severing how context can apply to other situations.	What other units of measurement combinations could the uncle say to describe his height? In groups of three, each of you come up with one value for a different unit of measurement that when totaled, equals the height of the uncle.	Expectation is on focused students engaging in the application of the task and concept.

To best equip students with skills that not only transcend surviving the day's lesson but become applicable life-long thinking habits, the emphasis is to foster the ability to provide conceptual explanations instead of calculational ones.

ELEMENTS OF MATHEMATICAL COMPREHENSION

Solving word problems is much easier when there is a command of numeracy, or the ability to use mathematical relationships to reason about numbers and numerical concepts. Rote memorization of basic facts, using mental mathematics, and understanding algorithms are all helpful tools as well. This list of skills could continue as I was not all-inclusive. For the purposes of conquering the word problem, I will focus on three key skills students need to comprehend mathematical problems:

- understand the relationship between quantities,
- embrace academic conversations to express and gain problem-solving perspectives, and
- differentiate between answering and making sense of the mathematics.

Visualization, an important link to connecting meaning, can occur with any or all of these key skills. To help students expand on their focus, visual representations are tools to reveal thinking and making sense of mathematics. Diagrams (also known as bar, strip, or tape diagrams) visually link the relationships between quantities as revealed in a word problem to uncover mathematical operations (Driscoll, Nikula, and DePiper 2016). Here is an example word problem in which students might be able to visualize how to solve before they are able to verbally explain how to solve.

Gabriel has ½ of an orange. He will give his half orange to Oscar, Soe, and Rosibel. If Gabriel gives each friend an equal piece of the orange, how much will each friend receive?

If Gabriel had several whole oranges, students may understand they will need to divide that whole number into three equal groups. However, dividing the fraction can be confusing. The students may visualize something like this to help them understand:

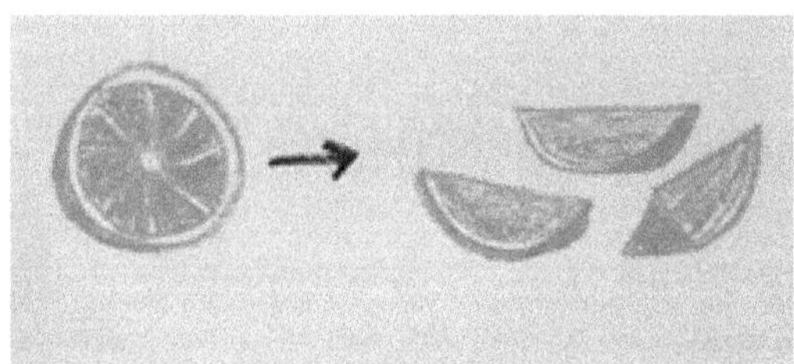

With guidance, this visualization translates into this diagram:

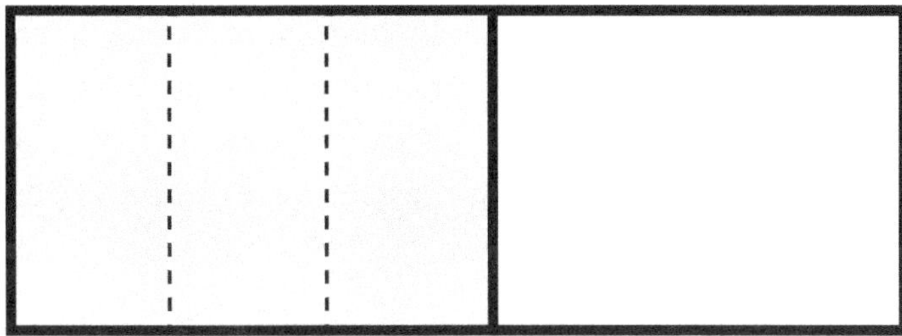

Through a diagram, students can identify relationships and make connections to operational meaning (Charles 2011). Diagrams are also a scaffold to applying language to mathematical information. In this example the diagram demonstrates three-thirds of one-half. Students can work together to build conceptual understanding of dividing fractions using visuals to link thought. For students who need support with quantitative analysis, expressing thought, and/or making sense of the mathematics, this approach can be impactful to attaining the concept.

Reflection Checkpoint
The calculational orientation approach would be to tell students that dividing fractions means multiplying the first fraction by the reciprocal of the second. How does the visual representation encourage a conceptual approach to learning how to divide fractions?

UNDERSTANDING THE RELATIONSHIP BETWEEN QUANTITIES

Characters in a story have identities and relationships. In a word problem, a number's identity is tied to its quantity and value, and numbers have relationships. Just like understanding characters in a story are more than just names, numbers in a word problem are more than just symbols. I remember in high school while reading *A Tale of Two Cities*, a classmate of mine kept making connections between characters that I couldn't see, and it frustrated me. I would have accused her of making everything up except the teacher applauded her observations. She had an acumen for character analysis. The equivalent of a character analysis in mathematics is a quantitative analysis. You can recognize your students that have a strong understanding of relating quantities by the ease in which they can translate word problems into expressions or equations. Here is an example situation I made up based on several years of informal observation you may find familiar in your classroom:

Teacher: Class, work with your partner to solve this problem together.

Quenton earned x dollars working on Monday. Allanah earned $100 less than twice what Quenton earned. Their gross income combined for $800. How much did Quenton earn Monday?

Student A: *thinking of where to begin.*

Student B: This is simple. $x + (2x - 100) = 800$

Student A: Huh?

Student B: So 4 can't be right, but 3 and 5 work because of 600 minus 100 so the answer is $300. This was easy.

Student A is obviously lost, and Student B believes she is explaining herself well by focusing on numbers. Student B internally performed a quantitative analysis and used reasoning to create the equation $x + (2x - 100) = 800$. She then shared aloud her thought process for solving for x without explaining how she came up with the equation. What Student A needed to understand was not how to solve the equation, but how to create it. The focus on numbers did not help Student A understand how Student B constructed the equation to solve for x.

The two may continue to talk about the word problem, but unless Student B explains her quantitative analysis and reasoning that led to forming her equation instead of focusing solely on the numbers to solve the problem, Student A will remain lost. Quantitative reasoning is imperative to writing the equation for what needs to be solved.

To understand a word problem, you must understand how the quantities and values relate to one another in a mathematical setting. A quantity is a measurable entity and the value of a quantity is the number—or the number and the unit—that represents the measured or counted quantity (Clement and Bernhard 2005). Here is an example word problem and one way to demonstrate a quantitative analysis:

Jessica uses 4 sweet potatoes for each sweet potato pie she bakes. She has 14 sweet potatoes. What is the greatest number of pies can she bake?

Quantity (measurable entity)	Value (number)
Number of sweet potatoes Jessica needs for a pie	4
Number of sweet potatoes Jessica has	14
Greatest number of pies Jessica can bake	Value I need

This table represents a quantitative analysis for the word problem. Quantitative reasoning occurred as you had to think of how the quantities and values 4 and 14 relate to one another.

Threshold of learning is the acquisition of concepts that transform perception and inspire analysis and exploration of the concept (Meyer et al. 2016). Solving for the number of pies Jessica can bake is most likely below your threshold of learning. When a task is below or above threshold of learning, disengagement appears in a variety of ways such as frustration or lack of motivation. You may not have found a motivational need to perform a quantitative analysis or identify your quantitative reasoning. Without quantitative reasoning, it might be easy to argue that using keywords and a focus on numbers to identify what operation to use is a way of performing quantitative reasoning. This places an incorrect emphasis on values (numbers) instead of relationships (Clement and Bernhard 2005).

The focus on numbers instead of relationships hinders further investigation and conceptual application. This becomes apparent with harder problems. Let us look at a trickier problem to challenge the keywords misconception and promote the need for quantitative analysis and reasoning. This was a conversation I had with my brother. Please keep in mind that my brothers and I grew up talking mathematics, constantly using quantitative reasoning to help us explain a point. Our conversation below was not uncommon. (I changed the values for mathematical purposes to give myself a raise, but the conversation and emotional turmoil stayed the same.)

Brother: Sister, I just got my quarter bonus, and it is twice what I expected!
Me: That's good. How much is it?
Brother: $40,000.
Me: [happy for my brother but trying to hide the sting that I can never make that much money in so little time as a teacher] Wow. Congrats, Bro. So, what you made in bonuses this quarter is two-thirds of what I make in a year. Are you sure the comma and decimal point are in the right places?
Brother: [silent pause] Yeah, we've been over this. What you make in a year I make in half-a-year. Sorry for bragging.
Me: Don't be sorry, Bro. I'm happy for you.

In contrast to presented word problems in a textbook, application, website, or worksheet, the ones that happen in conversation and everyday life are created and occur organically. Sifting through the information to find the unknown values means comprehending the *discussion*. In this example there was no final sentence to tell you what unknown value to solve for, but there were a few quantities without values. There is a way to make sense of mathematics through quantitative reasoning.

Performing a quantitative analysis focuses on quantities and their relationships rather than values. To understand the relationships between the quantities, you might have to reread the dialogue, establish what you know and what relationships are given to help you uncover what quantities and relationships to explore. As you revisit the conversation, I hope you are working in a learning community or with a coworker to perform a quantitative analysis so that maximum opportunities to gain perspective transpire. However, if you are reading this alone, there is still opportunity for growth, and either way I encourage you to fill out this empty chart before looking at my example. The number of spaces below does not indicate the number of measurable entities. If you discover more quantities, add them!

Quantity (measurable entity)	Value (number)

Here is an example quantitative analysis of the conversation:

Quantity (measurable entity)	Value (number)
Expected amount of the quarter bonus	One-half of the actual quarter bonus
Actual quarter bonus	$40,000
Fraction of the sister's annual income that represents the quarter bonus	two-thirds
Sister's annual income	Half of the brother's annual income without bonuses
Sister's quarter income	Unknown value
Ratio of the sister's income to the brother's	12:6 or 2:1
Brother's annual income without bonuses	Twice the sister's annual income

Brother's quarter income	Unknown value
Brother's quarter income with the expected quarter bonus	Unknown value
Brother's quarter income with the actual quarter bonus	Unknown value

This conversation does not include my or my brother's annual income because we both already know it. For those with a stronger literary background than mathematics one, our incomes might have been the only two unknown values you or your group contemplated. (Or you may have focused on the emotional aspect of my happiness combated by the sting of an educator's income which is *not* quantifiable.) However, when examining relationships between quantities, there is much more to discover. Notice how some quantities are combined with other quantities. For those with a stronger mathematics background than a literary one, you may have written even more quantities that I did not include in my example. You probably also found the values to every quantity because it was an itch you could scratch.

Once the understanding between quantities and relationships is established, exploration of different entry points for problem-solving commence. This focus on exploration of quantities and relationships is why Clement and Bernhard identify quantitative analysis as a more pedagogical approach to solving problems than finding keywords and a focus solely on values (2005). As exemplified in the conversation, a focus on values limits the mathematics and creates tunnel vision.

EMBRACING ACADEMIC CONVERSATIONS

Most everyone, within a gradient of difficulty, has struggled with a word problem. Academic language becomes paramount to convey a quantitative analysis, process thinking, questions, etc. Without language, drawing meaning and building on conceptual foundations is non-existent. Below is an extreme example of how some students with a learning disability, are emergent English learners, intimidated by mathematics, or reading below grade level view word problems. I created this exaggerated example purposefully to put you in the perspective of their possible struggles and to demonstrate the role of language in a word problem. It is a word problem written using Greek symbols.

Λυισ βουγητ 2 τιχκετσ το α χονχερτ. Ατ τηε τιμε οφ πυρχηασε, νινε ηυνδρεδ τηιρτψ-σεϖεν σεατσ ρεμαινεδ ιν αν αυδιτοριυμ οφ 10,000 σεατσ. Εξπρεσσ τηε αττενδανχε ατ α περχεντ οφ φιλλεδ χαπαχιτψ ατ τηε τιμε οφ Λυισ πυρχηασε.

The numbers—2 and 10,000—are easily identifiable, but the words do not aid in understanding. (Did you find the written number in the problem?) There is no question mark in this word problem. Those who are trained to restate the question might struggle with the concept that this problem does not seem to ask for anything.

Language is mandatory to explain this problem and what to solve. Language may be further needed to explain not only what the problem is requesting but for understanding the concept or to fill in gaps in knowledge. Try to solve this problem with a partner or in a group. What knowledge or concept did you gain through conversing that you would not have accomplished on your own?

There is an argument that students need to solve word problems alone because they will not have help from a peer on state-mandated tests. Learning cannot happen in isolation. Discussion amongst students fosters sharing of viewpoints and understanding. The advantage of multiple perspectives within structured academic discourse promotes "ways students can clarify their own thinking and consider and test other strategies to see if they are mathematically logical" (Parrish 2014, 11). Student-led academic discourse creates an environment where students provide the scaffolds and facilitate conceptualization (Walqui and van Lier 2010). Of course, the teacher provides guidance and addresses misconceptions as needed. For individual assessment purposes, students can solve problems individually once they have successfully gained problem-solving approaches from multiple perspectives.

Revisit the example written in Greek symbols. Perhaps in a group setting there is a cryptologist among you who can break the language barrier and explain the problem. You may need help forming the language to explain your thinking, in which sentence stems, frames, or vocabulary provide the support.

Once you understand the problem, performing a quantitative analysis can occur and you may find it easy to solve. This is another area of frustration for both students and teachers. How many times have you had a student that solves equations in isolation, but as soon as the mathematics is placed in a word problem, the student can no longer perform? Is this what just happened to you with this word problem written in Greek symbols? Perhaps you can calculate percent capacity if given the problem in isolation, but because it is written in a word problem, you no longer appear competent in this objective. Maybe your group did not figure out the word problem is about finding percent capacity. Perhaps you cannot read the written word, so you do not have a value you need to find the percent capacity. There are multiple avenues in which language became an obstacle that hindered your abilities. (For those of you who tried but could not decipher the word problem, I have included it in Appendix A to ease your mind.)

Words should not be a mathematician's enemy. Student engagement in indirect or implicit vocabulary acquisition of mathematics-specific and other academic words expands and improves their vocabulary knowledge (Vaughn and Linan-Thompson 2004). Had I known the academic phrase *quantitative analysis* the year I taught my student who always wanted to negotiate quantities and values in word problems, I would have been able to challenge him further because I would have understood the phrase and its application to quantitative reasoning. Language is necessary and should be encouraged to promote verbal metacognition so students can learn from one another while building a portfolio of academic vocabulary and language that no longer stigmatizes or inhibits acquiring mathematical concepts. Embracing academic conversations means talking about mathematics is a regular, perpetual occurrence even outside of mathematics class. It becomes a habit. It becomes normalized.

Reflection Checkpoint
How can you add mathematical conversations in the classroom outside of instruction?

MAKING SENSE OF THE MATHEMATICS

Differentiating between answering and making sense of the mathematics means understanding that the former solely requires calculating and the latter is understanding the purpose of the mathematics. In many ways, this is comparable to calculational orientation and conceptual orientation. Making sense of the mathematics leads to devising and/or calculating a solution. Thus, answering is *dependent* on the process of discovery.

This is a situation that would occur often in my lesson plans and implementation: On the first day I would introduce the concept and as a class explore the mathematics. The following day I would give the class more mathematics problems mixed with word problems that use the same mathematics concept. (The students are learning how to multiply by hundredths, so every word problem is about multiplying by hundredths.)

The word problems almost always revolved around the mathematics operations being taught. Consequently, my students did not have to focus on comprehending the words in a word problem as much as finding the values they needed to perform the operation they were learning at the moment. Their answers were *independent* of the process of discovery. There was no need for a quantitative analysis to make sense of the mathematics.

Skimming over the words to find the values to plug into the mathematics operation being taught that day was a survival strategy for my students who struggled in either reading or mathematics. It was also a quick way to get the work done for my students who were working below their threshold of learning. Students would get passing to high grades on their work, I would think they performed at an acceptable to mastery level, and we would move on to the next concept.

Martin notes that often correct responses are viewed as learning, and teachers are hesitant to explore a correct response which might disturb a student's confidence in the correct response (1996). However, by not implementing a culture of learning that explored making sense of the math, I paid a price. Although my students could easily solve multiplication word problems one week, some of them struggled desperately to solve the same type of word problems two weeks later when we had moved on to other concepts. Compound this isolated example with every possible word problem involved with all the standards students are responsible for understanding in a grade through performance-based assessments and it is no wonder some of my students would struggle with passing tests. With all the pressures teachers are placed under to teach a concept, differentiate it based on student learning needs and prior knowledge, maintain a high standard of rigor, and connect with every student's motivational need to sustain attention and desire to learn, it is easy to understand wanting to take teaching shortcuts. It is exhausting.

But then I remind myself of all the energy that would drain from me when a student who could answer a word problem one week could not answer a similar (sometimes the same) word problem later. This too, is exhausting. Making sense of mathematics is what promotes mathematicians' understanding as they apply and retain skills beyond one concept. By not implementing occasions to make sense of the mathematics during instruction, I had robbed my students of the opportunity to learn. The situation taught students that success in mathematics was not related to active sense-making and construction of ideas (Martin 1996).

EXPOSING THE MATHEMATICS

When comprehending a mathematical problem, the emphasis is to gain a mastery of the skills needed for conceptual understanding. Remember that visualizing occurs with any or all of the three skills:

- understand the relationship between quantities,
- embrace academic conversations to express and gain problem-solving perspectives, and
- differentiate between answering and making sense of the mathematics.

These three illustrated skills do not have a specified sequence. Some students may need to visualize the problem-solving process first. Others first analyze quantities to understand differentiating between answering and the process of solving. Other students may desire academic discourse and peer interaction from the start. Others still may use a different arrangement.

When comprehending a word problem, the learner must obtain conceptual understanding by interacting with text. This is intimidating for those who may not find comfort in reading about mathematics. I recently asked a coworker to help me with the language needed to solve a word problem and she groaned, stating, "I'm not that good at math, and I'm really not good with word problems." I prompted her to explain further, and she said, "The words hide the math."

Comprehending words and sentences will *expose* the mathematics. Fountas and Pinnell define reading comprehension as "actively working to construct meaning" that is "an ongoing *process* rather than simply the outcome or *product* of reading" (2006, *ix*). Because the goal of a word problem is to either solve for a value through devising an expression or equation, comprehending a word problem is the ongoing process of constructing the meaning of quantities and relationships to formulate an expression or equation to solve for unknown values. Transference and application of literacy skills *must* occur.

Addressing the reading in mathematics may not be an easy skill for educators whose pedagogical path focused heavily on mathematics and little on literacy. However, just like we cannot escape the existence of mathematics (even if we fail to recognize its presence) in everyday life, we cannot escape words. Word problems are the combination of two worlds that often attempt to avoid one another. For the sake of our students, educators cannot ignore this connection.

Sight words, or instant words, are the most commonly used words in our language. It is advantageous to internalize them automatically because of their frequency. Dr. Edward Fry developed an expanded list of the 1,000 most used words in publications, which he grouped in occurrence in increments of 100. His research states that 65% of published work is composed of the first 300 words (Fry 2000). When researching Fry's list of the 1,000 most frequently used words, I kept wondering if these same words occurred with the same frequency when applied to word problems. To glimpse into the frequency of words in word problems, I decided to pull data from a state mathematics assessment.

The following pages show data taken from the State of Texas Assessments of Academic Readiness (STAAR®) for third, fourth, and fifth-grade mathematics. I categorized every word and abbreviation—except the directions—for third, fourth, and fifth grade from the released mathematics STAAR for the year 2017. Because I only want to address a point, I purposely collected data for only one year for each grade. The following tables categorize each word on the assessment based on Fry's 1,000 instant words.

3rd Grade			
Fry's list of the 1,000 most used words in publication	Number of different words	Word count	Example word used in the assessment
100	67	680	one
200	34	146	answer
300	16	40	four
400	18	81	hundreds
500	22	77	equation
600	18	73	represent
700	8	20	fraction
800	12	37	equal
900	3	5	value
1000	7	31	total
Of the 124 Words not on Fry's list			
Proper nouns and abbreviations	37	79	Arkansas
Mathematics-specific	34	69	rectangle
Neither proper nouns, abbreviations, or mathematics-specific	53	166	crayons
Totals	329	1504	

Third-grade mathematics-specific words not in Fry's 1,000 most frequently used words list and how often they occurred:

centimeters	1	odd	1
comparison	1	parallelogram	1
cones	1	pentagons	1
congruent	1	perimeter	2
cylinders	1	plot	1
data	1	plus	2
diagram	1	prisms	2
digit	1	rectangle(s)	12
document	3	rectangular	2
dot	1	relationship	2
eighths	2	rhombus	1
equivalent	1	shade(d)	5
graph	3	strip	1
in.	4	thirty-five	1
labeled	1	trapezoid	1
minus	2	twenty-four	1
model	7	twice	1

4th Grade			
Fry's list of the 1,000 most used words in publication	Number of different words	Word count	Example word used in the assessment
100	60	510	two
200	27	83	show
300	25	76	group
400	19	52	measure
500	14	36	shape
600	12	40	difference
700	11	36	fraction
800	15	33	statement
900	8	24	term
1000	7	31	triangle
Of the 125 Words not on Fry's list (stem and leaf, for mathematical purposes, was counted as one word)			
Proper nouns and abbreviations	30	46	Wednesday
Mathematics-specific	41	96	polygon
Neither proper nouns, abbreviations, or mathematics-specific	54	123	neighborhood
Totals	323	1186	

Fourth-grade mathematics-specific words not in Fry's 1,000 most frequently used words list and how often they occurred:

acute	5	ounces	1
centimeter	2	oz	6
cm	4	parallel	1
comparison	1	perimeter	2
data	2	perpendicular	3
degree	1	plot	4
diagram	2	plus	2
digit	3	polygon(s)	6
displays	1	profit	1
document	3	protractor	1
dot	1	rectangle	1
estimate	1	rectangular	1
fl	6	relationship	2
fluid	1	segments	2
horizontal	1	shaded	1
kg.	1	stem and leaf	8
lb	4	symmetry	2
mass	3	vertical	1
minus	2	width	1
model(s)	1	zero	1
obtuse	4		

5th Grade			
Fry's list of the 1,000 most used words in publication	Number of different words	Word count	Example word used in the assessment
100	63	598	about
200	31	60	point
300	17	52	list
400	21	52	product
500	23	77	inch
600	14	36	length
700	10	23	section
800	15	37	symbol
900	10	20	weight
1000	7	16	bought
Of the 127 Words not on Fry's list (stem and leaf, for mathematical purposes, was counted as one word)			
Proper nouns and abbreviations	30	58	February
Mathematics- specific	49	137	coordinate
Neither proper nouns, abbreviations, or mathematics-specific	48	122	employer
Totals	338	1288	

Fifth-grade mathematics-specific words not in Fry's 1,000 most frequently used words list and how often they occurred:

centimeter	1	model	3
cm	2	ounces	2
combined	2	oz	4
comparison	1	parallelogram	2
composite	2	payroll	1
coordinate	3	pentagons	1
cube(s)	8	perimeter	3
cubic	3	plot	4
data	2	polygon	3
diagram	2	prime	1
dimensions	2	prism	4
dot	1	quadrilateral	2
equivalent	1	rectangle	2
ft	6	rectangular	5
graph	4	relationship	3
grid	1	rhombus	2
height	17	shaded	5
hundredth	1	stem and leaf	1
intersect	1	subtracted	1
kg	5	trapezoid	2

lb	4	volume	2
liters	1	width	1
M	3	x-axis	1
mass	5	y-axis	1
meters	3		

This is the data I collected when comparing Fry's list with mathematics-specific words not in the 1,000 most frequently read words.

Grade	Percentage of Fry's words in the assessment	Percentage of mathematics-specific words not on Fry's list of words in the assessment
3	79%	4.5%
4	78%	8%
5	75%	10.6%

Mathematics-specific words were only a small percentage of the words on the state assessments. Nonetheless, it is important to note that these words, if not read correctly or understood, can directly impact the students' ability to solve their respective word problems. Mathematics teachers must be purposeful in creating a print-rich environment in which students are exposed to the language of mathematics through listening, speaking, reading, and writing for acquisition and proficiency so that these content-specific words are also frequently used.

The word comprehension tasks that I will introduce to you focus on developing conceptual approaches to solving word problems. They include an example structure and vary in the sequential application of the three skills. It is important to note that students may work outside of the sequence of the structure and each task will need to be adjusted to meet your students' learning needs. Now that you have read some research and have a new perspective towards teaching word problem comprehension, give yourself time to reflect and then continue reading to explore some novel approaches!

Concepts Summary

- **Misconceptions about how to solve word problems stagnates comprehension.**
- **Teaching mathematics has two orientations: calculational and conceptual. Calculational orientation focuses on procedures and operations and conceptual orientation focuses on understanding and applying concepts.**
- **Comprehending word problems is an ongoing process of constructing meaning of quantities and relationships and forming expressions or equations to solve for values.**

CHAPTER 2

TAKING AWAY THE MATHEMATICS TO UNDERSTAND THE MATHEMATICS

I once searched TED.com for inspiring mathematics teacher speakers, and the teacher I found awed me and transformed my perspective. Dan Meyer, in his presentation *Mathematics Class Needs a Makeover*, argued that when we give problems to students, they are not involved in the formulation, the building, the process thinking of the mathematics in the world around them (2010). In his video, Meyer gives a few examples of how to facilitate student-led thinking around mathematics. He takes away the mathematics "to level the playing field of intuition" (Meyer 2010). It is a video that I encourage everyone to watch.

Thinking of my coworker who dreaded sifting through words just to find the mathematics, an activity in which the quantities and values are removed may seem like a backward approach. After all, a traditional classroom environment is "filled with students and adults who think of mathematics as rules and procedures to memorize without understanding the numerical relationships that provide the foundation for these rules" (Parrish 2014, 4). If I described your classroom, one in which students must memorize a set of procedures and use mnemonic devices to problem-solve, please don't think I am pointing fingers. I just described my own classroom for many years.

After trying Meyer's paradigm-shifting approach, the purpose of this student-led backward design to building and quantitative reasoning in a problem worked for a multitude of reasons. The approach fosters learning in response to inquiry and challenging situations within the learners' threshold of understanding. This is a characteristic of constructivism in which each person brings his or her own collection of knowledge to apply to the situation. It is here Meyer argues that removing the mathematics allows those with less mathematical knowledge an equitable opportunity for engagement. Leveling the playing field of intuition happens when a challenge occurs that prior knowledge alone cannot master. A "disequilibrium" arises that promotes learning, or the development of new or modified concepts as an adaptation (Martin 1996, 39). During this development, students build their conceptual knowledge of how mathematics fits in their understanding.

One major component to my success was that students intimidated by mathematical conversations now felt secure enough to contribute to the process. They did not feel a deficiency in mathematical discourse or concepts would prevent them from contributing to solving the problem. Students being able to build on one another's understanding empowered the group and perpetuated the safe environment for learning. Applying this idea of taking away the mathematics to promote mathematics is the first task I will introduce.

WHAT TAKING AWAY THE MATHEMATICS LOOKS LIKE AND WHY I INITIALLY FAILED

When I first tried this activity, it was not easy generating questions without adding quantities, relationships, and/or values. It took practice. Non-examples are just as important as examples to understanding this teaching concept. Here are some non-examples of questions that remove the mathematics and why:

My question	Why it does not promote student-led comprehension
It took us two class periods to finish half of this project. How long will it take us to finish the whole project?	In my question I gave students a quantity (the number of class periods) and a value (two) as well as what they need to solve. The students did not have to build a problem by discovering needed quantities or values themselves. They did not need to discover relationships. There was no intuitivc thinking.
It took Jonatan five minutes to return a book to the library. How can we turn this into a mathematics problem?	Students do not need to determine all their quantities or values because I gave some to them. Asking them to create a mathematics problem is counterproductive to organically working through a mathematics problem through collective problem-solving.

When asking a question, it is important to phrase it in a way that only presents a problem. Students need to build their understanding. They need to sort through what quantities and relationships are important and what the values are for each quantity. Here is an example list of questions I tried with students in my attempt to level the playing field:

- About how many books are in our library?
- How many pizzas do I need to buy for our pizza party?
- How many minutes have we spent reading *The Lightning Thief* by Rick Riordan?

Of the three examples, only one question was met with resistance and frustration. Can you guess which one and why?

I had failed the first time I tried to remove the mathematics, but I was used to learning from my failures. If you guessed the library books question was the one that failed, you are correct. I took my third-grade students to the school library, told them to look around, and asked them "About how many books do you think are in our library?" I had gotten the librarian's blessing to use our thirty-minute library time working through this mathematics problem. That half-hour was a huge mess. I am glad I did not quit when my students could not (collectively or individually) problem-solve the estimated number of books in our library. Some students immediately started counting each book, one-by-one. Others complained that there were too many books to count and there was not enough time. Some came close and talked about multiplying the number of bookshelves times another number, but they couldn't agree on what the other number would be. It was good that they understood the relationship between the bookshelves and total number of books, but they struggled with finding the value of their second quantity because they did not know what their second quantity was and had no academic vocabulary to express it. I still remember the librarian's pity-smile when she told us the exact number of books before we left, defeated. Upon reflection, I think asking my students to estimate the number of books in our library was a little overwhelming for them. How could I have done a better job setting them up for success? At the time I didn't know.

After some experience I have gotten over the embarrassment of that moment. The students were not as strong in building upon one another's knowledge because they were not used to communicating and problem-solving in the situation and environment in which I had placed them. They needed scaffolded support with language—expressing what they needed to know—and most importantly, I needed to set them up for success with a much smaller and more invested task before overwhelming them with calculating an estimated 10,000 books.

Additionally, the task did not have a personal value to them that calculating the number of pizzas I needed to buy for them did, or the prideful investment that only finishing an 87,223-word book could muster. (The latter question came towards the end of the year.) When I asked the students how many pizzas I needed to buy for our party, some of the quantities included: the number of slices each person would eat, the number of students in the class, how many different pizzas they would want, and if they included their teacher. They collectively sifted through all the quantities to determine which ones were needed to answer my question and assigned their values. During this process they discarded quantities that were not important, such as the number of the types of pizza to buy. After looking at values, they realized that there was a relationship between quantities missing "as a result of making sense of the situation's quantitative structure" (Clement and Bernhard 2005, 363). They needed to know the number of slices in one pizza. On the inside I was jumping for joy. It was a beautiful teaching moment. Had I given up after the embarrassing library incident, I would have robbed my students of the many opportunities they had to generate their own approach to mathematical problem-solving.

APPLYING A NEW CONCEPT: TEACHER DIFFERENTIATED SUPPORT

This activity might be a new concept for you. Sometimes it can be difficult to apply a new concept. Here are some examples that you can try in your classroom to get started. The questions range in the complexity of quantities and relationships, so you can pick and choose the ones you are most comfortable exploring with your students.

Example question	Example of possible quantities (not all-inclusive)	Example of possible relationships between quantities
How many - do I need for each student?	The number of students The number of - for each student	The relationship between the number of students and the number of–
How much time do we need to finish this project?	Amount of time for that content block Amount of work	The relationship between how much work is done per content block
How many bottles of water do we need for our field trip?	Number of students Number of bottles of water each student needs	The relationship between the number of students and how much water they will consume

About how many books are in our library?	Number of bookshelves Number of books on a bookshelf	The relationship between the number of books on a bookshelf and the number of bookshelves
How many pizzas did I buy for our pizza party? (Do not let students see the number of pizza boxes.)	Number of students Number of slices in a pizza Number of slices each student ate Number of slices remaining	The relationship of the total number of slices eaten and how many slices are in a pizza The relationship of the total number of slices remaining and the total number of slices eaten
How long will it take us to finish this book?	The number of pages in the book The number of pages read in a content block	The relationship between the number of pages in the book and the amount of time it takes to read a page

Reflection Checkpoint
What is a question you can pose that is relevant to your students? How can this question spark a disequilibrium to promote learning?

THE CONNECTION TO COMPREHENDING WORD PROBLEMS

Taking away the mathematics not only levels the field of intuition, it also prompts construction of ideas by the individual and group. These ideas are absorbed into the students' collection of knowledge. It is this active moment of sense-making in which the teacher gains access to student thinking, clarifies misconceptions, and guides further application. It is also the teachable moment in which a connection to word problem comprehension happens. Just like reading teachers should not take the joy from reading by making students dissect every word of a book, mathematics teachers should not take away from the joy of organic problem-solving by requiring students to write every word they said. Knowing they would have to write down everything they say might inhibit some students from speaking their mind. I wrote down the question and took notes of their verbal metacognition of how they solved these "intuitive" questions. I gave some of their thinking academic names (such as quantity, value, relationship, represent, unknown, etc.) to empower their discourse. The written process became an anchor chart hung visibly in the classroom. Here is an example:

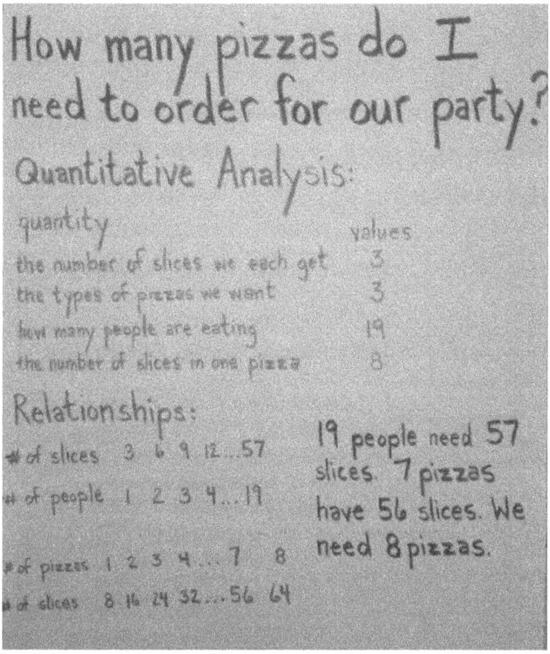

You might look at my example chart as an exemplar, or you might be quick to notice I did not label the diagrams or some other fault. Please note that anchor charts meet the need of the students at that point in time. For example, if diagrams need to be in a lined table, then demonstrate this visualization. It is up to you how you wish to facilitate concepts based on your students' collaborative building on mathematical perspectives. The purpose of the anchor chart is to work within your students' understanding to scaffold their thinking, honor their voice and ownership in their learning, and give them a foundation from which to build their metacognition and knowledge. Additionally, there should be more than one anchor chart to model varying perspectives and different approaches. They become print-rich resources of the mathematics world that students reference often while visualizing and working through:

- quantitative analysis and reasoning,
- verbalizing metacognition to share and gain perspective, and
- differentiating between answering and making sense of the mathematics.

Building on their shared, invested connection to mathematics, students can embrace the academic language in a print-rich environment. This helps them embrace word problems with a more level playing field of conscious mathematical reasoning. The exposure to language and academic vocabulary normalizes word problems and helps students realize that words do not hide the mathematics. The ownership and exposure to constructing quantities, values and determining operations prepares them for deciphering word problems. It equips them with expressing their thinking, building knowledge in a collaborative environment, and application of acquired concepts. All of which are life-long skills applicable beyond mathematics.

Concepts Summary

- **Students intimidated by mathematics can engage in mathematical conversations when the field of intuition is leveled. This means posing a problem in which students collectively build their own quantities and values and use reasoning to determine the operations.**
- **The role of the teacher is to encourage student-led construction of ideas, address misconceptions if the group does not, and record their thinking.**
- **Use anchor charts to visualize and name thinking based on student discourse to empower student voice. These anchor charts become a foundation to comprehending word problems.**

CHAPTER 3

FOCUSING ON QUANTITIES, RELATIONSHIPS, AND VALUES

The way I structured my mathematics block in the latter years of my teaching career was to first teach the class whole group. Then, as they worked together or individually, I made myself available at the teacher table for anyone with questions. I would get one or two regulars at my table, but for the most part, my students that needed additional support varied based on the concept. I had a student who was proficient with solving equations, but when he had to solve a word problem he always came to my table. Together with other students seated at my table we would read the word problem twice, underline keywords, circle important numbers, and rephrase the question. He was robotic during this process, as so many of my students were, and didn't need my guidance to go through the steps.

What made him different from my other students that struggled with word problems was how he behaved after we finished the process. He would *negotiate* the numbers and what quantities needed values with me with hypothetical questions like, "But Ms. Kue, what if it was unequal groups instead of equal?" If I had understood the discourse he wanted revolved around quantitative analysis and reasoning, I would have been able to give a name to what we were doing, and an identity to his thinking. Unfortunately, I was not in that place of my pedagogical journey. Instead, I acknowledged I had a student looking for a challenge and engaged in negotiating numbers (values) and what needed numbers (quantities) with him to change outcomes. He could not thrive in the calculational orientation learning environment and pushed me to promote a conceptual orientation with his need to explore relationships in a word problem. I reflected on how this gifted student needed to learn and added a new approach to my teaching the following year.

This next word problem comprehension task is the inverse of the previous one. Instead of posing a question and eliminating all quantities and values, this activity requires a focus on differentiating between given quantities and values and forming relationships to create a problem. The focus on numerical relationships perpetuates the purpose of conceptual orientation.

The procedure for this one is simple:

1. Take away the final sentence of a word problem.
2. Students perform a quantitative analysis and create a final sentence that requires at least one operation or asks for an expression.

I first tried this activity when I was still a classroom teacher. As I mentioned, I was unfamiliar with the term and concept of quantitative analysis and quantitative reasoning and just thought I had created a novel approach to problem-solving. My students enthusiastically wrote the last sentences to their word problems and I remember thinking, "Do they not realize they are giving themselves work to do?" I tried to comprehend what they got from the task with little success. However, I noticed my students that could not write a final sentence were the same ones that struggled with solving word problems on tests. The task became a way I informally assessed word problem comprehension and who would receive interventions.

After transitioning from teacher to writing curriculum and providing classroom instructional support, I began to explore why my old students enjoyed the task, why some succeeded and some struggled. My informal research suggested students that could create the final sentence of a word problem also understood the relationship between the quantities. We are now going to revisit performing a quantitative analysis. After making the analysis, quantitative reasoning is applied to identify and create relationships (Charles 2011). Here is the example:

First, begin with reading the word problem. I purposely left out the final sentence.

The school bus makes three stops every morning before arriving at school. At the first stop, 12 students get on the bus. At the second stop, half as many students as the first stop get on the bus. When the bus arrives at school, 32 students exit.

Next, record all quantities. For each quantity, think, "*What is the value of this quantity?*"

Quantity	Value
Number of stops the bus makes	Could be information used when solving the problem—3
Number of students at the first stop	Could be needed to solve the problem—12
Number of students at the second stop	Could be needed to solve the problem—half of 12
Number of students exiting the bus	Could be needed to solve the problem—32

Finally, apply quantitative reasoning to discuss what new quantity can be explored and write a sentence to solve for the value of this quantity.

You may have noticed I left out the quantity of the students at the third stop and its value. Students with a foundation in quantitative analysis and reasoning will immediately notice the numbers don't add up, and create an equation to solve for a value to match the quantity. Others with a stronger literary focus will read that there are *three* stops, but only *two* are mentioned, and look for a value for the third quantity. Students whose strengths lie elsewhere may feel overwhelmed with the whole task. Thus, requiring students to think through this process and record their final sentence individually robs them of the opportunity to learn from one another.

Here is a quantitative analysis with the same word problem with the last sentence.

The school bus makes three stops every morning before arriving at school. At the first stop, 12 students get on the bus. At the second stop, half as many students as the first stop get on the bus. When the bus arrives to school, 32 students exit. How many students were added to the bus's student total at the third bus stop?

Quantity	Value
Number of stops the bus makes	Could be information used when solving the problem—3
Number of students at the first stop	Needed to solve the problem—12
Number of students at the second stop	Needed to solve the problem—half of 12
Number of students exiting the bus	Needed to solve the problem—32
Number of students at the third stop	Value I need

SOCIAL LEARNING AS DIFFERENTIATED SUPPORT

Consider this example of how collective exploration improved my understanding of unknown values and how to solve for them. I conversed with two high school juniors before school on the complications of word problems and one student said this, "Word problems are easier for me to solve than when I am given just an equation." Her statement left my mind blank, and I prompted her to explain. "I can solve a problem if the equation is given to me but I find it hard to solve a problem with only instructions to evaluate it because I don't remember from class how to solve it or I don't have any idea how to solve it. For example, if someone gave me a piecewise function, I would like the word problem more because it gives names to variables and points out what variable I need to isolate. I have a better idea what I'm solving for when the word problem gives me more of a background of what my next step would be based on wording because I know the goal."

What a contrast to my coworker who stated, "The words hide the math!" I wonder, if the two of them worked together on word problems, what academic discourse would occur, and what portfolio of scaffolded supports, reasoning, and effective approaches they would compile over time.

Social learning is imperative to filling gaps in comprehending, challenging misconceptions, and adding perspectives to problem-solving. It may be that the student who can perform a quantitative analysis with ease struggles with explaining thinking. The student who comprehends there is a missing quantity might not always know how to solve for the value. The student who can visualize the problem may help others connect relationships. The student who comprehends neither quantity nor value and struggles picturing the problem gains all these viewpoints through engaged academic discourse. While questioning, visualizing, and explaining the problem with one another, students sharpen their ability to verbalize their reasoning, thought process, and perspectives. Consequently, they also clarify their own reasoning, investigate and apply other approaches, and make decisions about which strategies prove most efficient for specific problems, building a "repertoire of efficient strategies" (Parrish 2014, 11). This engaged, peer scaffolding supports each student as the group customizes their development for this problem while they engage in quantitative reasoning.

As a component to social learning, academic discourse is imperative but may need scaffolded language support. Here are some example sentence stems students may use:

- One quantity is...
- The value of that quantity is... and is needed/not needed to solve the problem because...
- The value I/we need is... because...

I have included some thought prompts and other response stems to help with expressing in Appendix B. I want to stress using the word *because* when students share their thinking. When they say *because* they explain their metacognition. This means they must first reflect on their own understanding before processing how they can convey their ideas. It is a powerful tool for effective communication because it can both strengthen understanding and/or address misconceptions.

I wanted to share this strategic approach with other teachers during a professional learning session. The problem was one I took from a second-grade classroom. It was something like this:

Allen has collected 15 apples. Otis has collected 20 apples.

To demonstrate exploring the quantities and values, I asked them to write a final sentence that required at least one operation. Even though I encouraged them to talk to one another, the task was below their threshold of learning and they quickly jotted down a final sentence without speaking. Some didn't even bother to write a final sentence. I would not be surprised if you thought of something similar to their responses, which was either to add or subtract the apple amounts.

- How many apples do Allen and Otis have all together?
- How many more apples does Otis have than Allen?

Here is a quantitative analysis for each final sentence:

Quantity	Value
Number of apples Allen has	Needed to solve the problem—15
Number of apples Otis has	Needed to solve the problem—20
Total number of apples	Value I need

Quantity	Value
Number of apples Allen has	Needed to solve the problem—15
Number of apples Otis has	Needed to solve the problem—20
Greater amount of apples Otis has than Allen	Value I need

But there was so much more to this exercise, and after securing their foundational understanding, I prompted a deeper conceptual orientation. “I noticed none of your final sentences added new numerical information. Let’s be consistent and make this a rule. Please write two more sentences without adding additional numerical information. One will require multiplication; one will require division.”

Here are the two sentences again, so that you may try to create a multiplication and division word problem:

Allen has collected 15 apples. Otis has collected 20 apples.

I was met with silence. The teachers were either hesitant to talk, were not used to processing aloud with a partner, or did not know how to start. I encouraged them to talk with their group and slowly the volume of the room rose. The responses were all similar to the following:

- How many apples do Allen and Otis have when you multiply them together?
- How many apples do Allen and Otis have when you divide them?

Here is a quantitative analysis for each final sentence:

Quantity	Value
Number of apples Allen has	Needed to solve the problem—15
Number of apples Otis has	Needed to solve the problem—20
	Value I need

Quantity	Value
Number of apples Allen has	Needed to solve the problem—15
Number of apples Otis has	Needed to solve the problem—20
	Value I need

I could not think of how to explain the quantity for either of these final sentences. However, there is still a value needed. To find one value, I multiply the two numbers. To find the other value, I divide the two numbers. What is the quantity? What is the relationship between these quantities which creates an unknown value?

Reflection Checkpoint
What is your final sentence that requires multiplication or division?

Unlike the addition and subtraction questions the group had generated, neither of their new sentences explored the relationship between the quantities. Consequently, there was no quantitative reasoning. I know the teachers struggled with creating just one sentence because some of them even prefaced that their sentences were "not very good" before sharing. The groups' frustrations were growing by my asking if the final sentences created word problems that required a quantitative analysis or reasoning. One teacher commented loudly that the two sentences aren't natural for multiplication or division.

To acknowledge her (and all the participants') struggle, I commented that this was a difficult task and that it would feel unnatural to shift perspectives. Another teacher asked me how I would have written my sentences, and I showed them the following sentences:

- If Allen and Otis kept collecting apples in the same increments as their current totals, how many times will each person collect apples before they have the same amount?
- What is the percent of Allen's total apples to Otis's total?

Here is a quantitative analysis for each final sentence:

Quantity	Value
Number of apples Allen has collected in one increment	Needed to solve the problem—15
Number of apples Otis has collected in one increment	Needed to solve the problem—20
Number of apples Allen and Otis collect that is the same	An unknown value I need to solve the problem
Number of times Allen collects apples to get to the shared number	Unknown value I need
Number of times Otis collects apples to get to the shared number	Unknown value I need

Quantity	Value
Number of apples Allen has	Needed to solve the problem—15
Number of apples Otis has	Needed to solve the problem—20
Percentage of Allen's apples to Otis's total	Unknown value I need

WHAT WORDS REVEAL

I would be a liar if I did not admit that I hope many of you paused and reflected on my two final sentences. Notice that although I did not add numbers to the multiplication problem, my final sentence established new quantities and relationships that need values integral to solving the problem. This strategic approach of constructing a final sentence acknowledges that words can be used to explore numerical development, and perspectives can shift as their relationships change, much like character development in a novel. While I am on the topic of literacy, let me reemphasize the focus on text as a base for comprehending the situation and finding values.

I had been instructed that my students should read each word problem at least twice to ensure students understood the problem. Reading the problem twice did not guarantee students understood the relationship between the quantities and values. As an educator, reading "Allen has collected 15 apples. Otis has collected 20 apples" is below your threshold of understanding. You do not need to read these sentences twice to understand them. However, how many times did you reread these sentences or repeat them aloud because developing a relationship between the quantities *was* within your threshold of understanding? The task shifted from understanding the words to understanding what the words disclosed. Words have significance in a word problem. Students must understand that often they reveal quantities, values, and/or relationships.

Here is a more challenging word problem with more quantities and values to explore:

> Fifteen soccer players are selling various chocolate bars for a fundraiser for their team. Coach Cindy ordered the following amounts of chocolate:
>
> - Four cartons of chocolate bars with 20 bars in each carton
> - Six cartons of chocolate almond bars with 10 bars in each carton

Are you able to create a final sentence that requires a quantitative analysis and at least one operation?

If you are reading this with others as a form of cultivating or perpetuating a learning community, then you already have the resources for social learning. Your group will culture your own zone of proximal development as you verbalize your metacognition. Notice how many times you read with a focus during the process of making sense of the mathematics. If you implemented a quantitative analysis for the chocolate question alone, then I suggest you find a colleague or someone with a similar interest to form his or her own quantitative analysis. The two of you can share your final sentences to discuss and benefit from the language and perspective of one another. Connections to prior knowledge, explicit use of academic vocabulary, and application of acquired knowledge enhance learning and increase through discourse. Zone of proximal development expands. Growth and understanding of a concept demand perspective; do not rob yourself of this conceptual orientation or the life-skills it cultivates.

AN ISSUE WITH MNEMONICS

Teaching mathematics varies based on the teacher's conceptual understanding of the content objective and mathematical pedagogy. I want to revisit conceptual orientation and its value in contrast to calculational orientation. Recall that conceptual orientation explores the how and why of mathematics while calculational orientation is about solving the mathematics in isolation. It is hard not to notice the various anchor charts that address problem-solving strategies based on calculational orientation in a mathematics classroom. You may even have one hanging in your classroom. I know I did when I was a teacher. Here is an example anchor chart I see often known as CUBES:

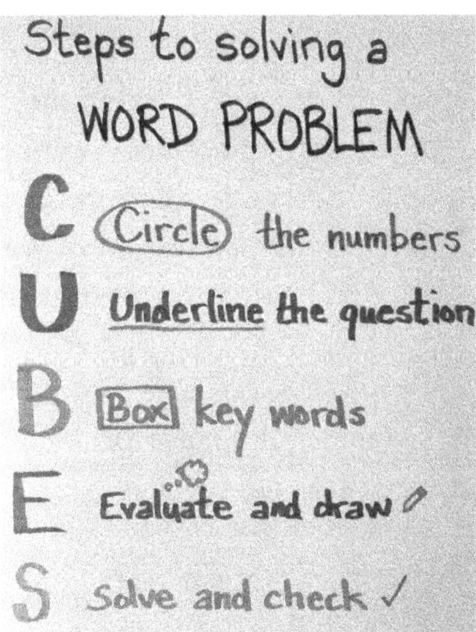

When I was a classroom teacher, I wondered what was missing from this strategy. My students who struggled the most with word problems and faithfully followed CUBES still did not fare better on assessments. It was frustrating because they were working harder than ever, writing more than was necessary for each word problem, and yielding similar results. As I researched, I was able to clarify what I observed. The first epiphany is simple—where does comprehension of the text appear in this acronym? Circle, underline, box, eliminate, and solve are not synonymous to the verb *comprehend*, and this is an important action to leave out of a strategy for solving word problems.

Secondly, circling numbers is not a quantitative assessment that empowers quantitative reasoning. Neither is boxing keywords. In problem-solving strategies, there is often a focus on the numbers in the word problem when the focus should be on *quantities*. Further evidence of the importance of quantities is in the word problem that gives extra information. Here is an example in which the problem cannot be solved correctly without understanding the relationship between quantities:

> Enzo, Ian, Uri, and Kael are running laps. Enzo ran 12 laps. Ian ran twice as many laps as Enzo and 2 fewer than Uri. Kael ran 3 times as many laps as Uri. How many laps did Uri run?

The relationship between the number of laps Kael ran and the number of laps Uri ran is not important to solving this problem. It is an unnecessary quantity. Therefore, the value is unnecessary. The person solving this problem cannot correctly determine which values to use without understanding which quantities (and their relationships) are important. Circling *three* and boxing *times* would mislead the student on how to solve the problem.

Finally, and it is not implicit, but learners should be able to work together to evaluate, visualize, and solve problems so they can build their repertoire of efficient strategies.

Concepts Summary

- Creating the last sentence of a word problem gives students the opportunity to construct meaning through quantitative reasoning.
- When students converse with one another they fill in gaps in learning, challenge misconceptions, and gain problem-solving perspectives.
- Problem-solving strategies such as mnemonics do not address comprehension or quantitative reasoning.

CHAPTER 4

JIGSAW WORD PROBLEMS

The purpose of jigsaw reading, as it is used in Reading, Language Arts, or Literacy-based instruction, is to create a cooperative learning environment in which the assigned task cannot be completed without each member's contribution. For example, a group of four students may each be tasked to read a different portion of a significant historical figure's life. Knowledgeable of only their portion, the students come together and share what they learned to create a complete picture of the historical figure's life. Of course, this is only one *brief* example of jigsaw reading. There are varied forms and structures applied to this idea. When executed correctly, the concept is condensed to this notion: students must share their understanding and place value on each other's knowledge because no task can be completed without everyone working together. The outcome is a learner-centered activity in which students read, listen, and speak (and write, depending on the assignment) with one another to piece together their information and form a larger understanding. This is all good stuff.

Why not extrapolate and transfer the jigsaw concept to word problems? Why not cultivate reading with a mathematical purpose to create mathematical readers? Literacy teachers shouldn't hog all the strategic approaches to purposeful comprehension instruction!

So how does this work for a word problem? Here is an example word problem:

Five friends spent $105 at the movies. They spent $50 on tickets and $21.25 on drinks. The rest of their money was spent on snacks. What is an equation that can be used to find *s*, the amount of money in dollars the friends spent on snacks?

Here is one way you can structure jigsaw word problems in your classroom:

1. Rewrite the word problem so that each sentence is isolated on its own sentence strip.
2. Students sit in groups of three (if the problem is three sentences long) or four (if the problem is four sentences long).
3. Place the sentence strips face down in the middle of the group and let each student choose a sentence strip.
4. Students do not share with their group but may ask for assistance as needed from an outside party while reading their sentence.
5. After everyone in the group has read their sentence to themselves, they begin piecing together their problem. The person who believes he/she has the first sentence of the problem says, "I think I have the first sentence of the problem because..."
    a. Note: When students justify their answers, they do *not* read their sentence aloud. This makes them actively justify their reasoning instead of passively reading.
    b. Example: *I think I have the first sentence of the problem because my sentence gives me a whole picture. It tells me how many friends there are and how much money they spent.*
    c. Non-example: *I think I have the first sentence because it says five friends spent $105 at the movies.*
6. If everyone in the group agrees that the student does have the first sentence, the student then reads the actual sentence aloud to the group and places it in the middle of the table.
    a. If someone else in the group believes he/she has the first sentence, that person then says, "I think I have the

first sentence of the problem because..."

   b. The group, after listening to the two explanations, decides which person has the first sentence. That person then reads the actual sentence aloud and places it in the middle of the group for all to see.

7. The person who believes he/she has the second sentence then says, "I think I have the second sentence because..."
   a. Example: *I think I have the second sentence because the first sentence mentions how much money they have, and my sentence mentions how much money they spent. (Now I know "they" is friends, and they are at the movies.)*
   b. Non-example: *I think I have the second sentence because it says they spent $50 on tickets and $21.25 on drinks.*
   c. Follow the same procedure for step 6. After reading the actual sentence aloud, the student will place the sentence to the right of the first sentence.
8. The student who thinks he/she has the third sentence says, "I think I have the third sentence because..."
   a. Example: *I think I have the third sentence because the first two sentences mention a whole amount and partially spent amounts. My sentence mentions how the rest of the money is spent. It is another partial amount.*
   b. Non-example: *I think I have the third sentence because they spent the rest of their money on snacks.*
   c. Follow the same procedure for step 6. After reading the actual sentence aloud, the student will place the sentence to the right of the second sentence.
9. The student with the remaining sentence says, "I think I have the fourth sentence because..."
   a. Example: *I think I have the fourth sentence because it asks for an equation to solve a missing amount.*
   b. Non-example: *I think I have the fourth sentence because it says, "What is an equation that can be used to find* s*, the amount of money in dollars the friends*

spent on snacks?"

c. The student then reads the actual sentence aloud and then places it to the right of the third sentence.
d. If the group decides that their problem is out of order, they may work together to rearrange it. However, this should be a group task and not a decision dominated by one or two students.

10. The students then engage in academic discourse about how to solve the problem. They record their answer.

Here is the problem again, when broken into individual sentence strips:

Problem 1 Five friends spent $105 at the movies.
Problem 1 They spent $50 on tickets and $21.25 on drinks.
Problem 1 The rest of their money was spent on snacks.
Problem 1 What is an equation that can be used to find *s*, the amount of money in dollars the friends spent on snacks?

You may have noticed the "Problem 1" typed on the right of each strip. When I use this activity with students, I often have more than one problem to piece together, and it helps me differentiate the problems to minimize confusion. For example, one group may have Problem 1 while another group may have Problem 2. And while we are on the topic of differentiation...

Some suggestions for differentiation:

For the student who needs reading support:	For the student who needs support with the problem-solving:	For the student who needs something to struggle over:
Create a safe environment in which students can go to you for help reading their sentence before returning to their group. You might say, "As you read your sentences to yourself, I will be at the teacher table for anyone who might have some words they don't know or understand."	Place this student with other students who are strong in verbalizing the problem-solving process. Keep thought phrases in a visible area that state "this sentence introduces the whole, this sentence mentions equal groups, this sentence mentions part of a whole, this sentence is asking for a part of a whole" etc. to guide language for expressing metacognition.	Create problems that require a higher application for them to solve. The solving process or explanation may require a new conceptual approach or new academic vocabulary. Give a group of students an image, such as a grocery store aisle, and tell them they will need to write a word problem. Each student will need to determine which sentence to write, and the combined sentences become the word problem.

For all students:
Before students begin piecing together their sentences in their group, have all students with the same sentence come together and talk about their sentence. For example, say, "If your sentence starts with the word five, come to this corner of the room." As each group reads their sentence and collectively thinks of how to express where their sentence belongs, students who need help with reading get to hear their sentence, and students who need help expressing get to hear how others express. This "brainstorming" gives students a resource to use when they go back to their groups and make their arguments.
Give students scratch paper to record or illustrate their thinking.

WHY I INITIALLY FAILED

The first time I thought up this activity, I was in a professional learning session about teaching how to comprehend nonfiction texts. At the time, I was an educator tasked with writing math curriculum. I am sure some of you can relate when I mention that I had no choice in attending a session that was not my content. I struggled with the whole session, trying to translate the literacy strategies into applicable mathematical approaches. While the instructor had us jigsaw read a nonfiction text about a hippopotamus, another mathematics teacher had reached the end of her patience and said in a loud, disgruntled voice, "And how do I use this jigsaw strategy when I am teaching students about fractions?" A few of the other unhappy mathematics teachers sprinkled around the room mumbled in agreement. The instructor stood before us, flustered, and gave a weak response about how there is reading in mathematics as well. It did *not* appease the crowd.

This moment turned into my epiphany, and the idea to jigsaw word problems was imagined. I found myself excited as I wrote out the process and could not wait to add this strategic approach to problem-solving to my instruction. When I eagerly introduced this activity to a group of students, I failed. I tried again the next day, and it was even more disordered and counterproductive to my end goal. It was one of the biggest failed implementations of my teaching career, and even though I liked my idea, I decided that I was wasting students' time. The state test was just a month away. I felt the pressure to get them test-ready and trying something so different was too much of a risk. I returned to my old self and supported the classroom teacher's calculational approach: reminding students to identify keywords, underline the question, and rewrite what the question was asking. It was much safer, and they were not as "lost" as they had been those two days I went rogue.

Nevertheless, I remained bothered that this approach to problem-solving was not successful. On paper and in theory it was structured, purposeful, and revolved around student-centered learning. That summer I spent hours contemplating if the activity was worth another try and convinced myself that I had not modeled the structure of the task properly.

The example problem was what I used with this new group of fifth-grade students. They had been working with algebraic reasoning and the word problem worked with the concept. Although I did not record the events of my first colossal fail, I did jot notes the next time I introduced this activity for the first time to a group of students. According to my notes, this happened:

Student 1: "I have the first sentence because it says, 'They spent $50 on tickets and $21.25 on drinks.'"

Group: "I agree."

- "Yeah."
- "I agree too."

Student 1 placed the sentence down and everyone quietly looked at it. [I am not sure if anyone read it.]

Student 2: "I think I have the second sentence, but I don't know why."

Student 3: "Okay, put it down, and let us see it."

Student 2 placed her sentence down on the desk in front of her. The rest of the group read her sentence and then nodded silently.

Student 3: "I know you [points to Student 4] are next because my sentence ends in a question mark, so I have the last sentence."

Student 4: "Okay then. I have the next sentence because... it says they spend money on snacks."

Student 3 started talking before Student 4 even placed his sentence down for everyone to read. "I have the last sentence because like I said, it has the question mark."

The students then looked at the sentences on their desks and moved them around before Student 3 told the group they would have to add $105, $50, and $21.25 together for their answer.

My instructional execution was still unsuccessful. Students were not comprehending word problems through an ongoing process of analyzing each sentence and how it pieced together based on exposed quantities. Not only was there no academic discourse, students lacked metacognition and only one voice (incorrectly) told the group how to solve the problem. I had failed yet again.

Fortunately, I had taken notes. I re-read the student interactions and reflected on how I presented the task. I realized that the students were not familiar with this approach to processing, and that it would take more time, modeling, scaffolding, and social interaction to yield results. I understood that success demanded a better way to foster student-to-student interaction and scaffolding but I did not have the words to explain why or how. While researching for this book I discovered the answer, and it involves perspective. I focused on how to get students to understand the quantitative reasoning within the word problems, but I was missing the foundation of reading comprehension. "Responsiveness is critical for understanding" (Beer and Probst 2017, 29). Students were collecting information based on their sentence without thinking of its role in a problem because they were not responsive readers. The responsive reader tries to make sense of the text and is responsive to the thoughts and perceptions of *other* responsive readers (Beers and Probst 2017). To become successful at teaching responsiveness, I would need to prioritize the focus on text, so students could process how to use the words to expose the relationship between numbers.

How much would change if I let students go out of sequence when putting together their word problems? Every time I did this task with students, the one with the sentence ending in a question mark always said something along the lines of "Yes! I got the last sentence!" If all the students know it is the final sentence, why not let them respond to what they know?

I tried again at the same school and grade but with a different set of students and with the same structure to the task *except* for the part where they could only argue for and present their sentence in order. I also instructed, "If you say you have the last sentence because it ends in a question mark, you also have to let everyone know what the question wants solved." With three student volunteers, we modeled the process in front of the class. Finally, I gave the groups the word problem.

Student 1: "I have the last sentence because it ends in a question mark. It wants us to write an equation for how much money some friends spent." He put his sentence down and read it out loud for everyone.

Student 2: "Okay, then I have the first sentence because it says what they spend their money on." He put his sentence down, and while he was reading it, Student 3 interrupted.

Student 3: "Wait! I think my sentence is first because your sentence says 'they' and we don't know who 'they' is."

Student 2: "What do you mean?"

Student 3: "I mean, who is *they*?"

Here the group sat in silence, and I prompted them. "If you did not know the last sentence, would you know that the pronoun 'they' was talking about friends?"

The group: "No."

Student 3: "That's the point I am trying to make. We don't know who 'they' are yet."

Student 1: "Because we already know they're friends from my sentence?"

Student 3: "Exactly."

Student 2: "Okay, whatever."

Student 3: "I think I have the first sentence because it talks about five friends going to the movies and how much they spent." All students agreed. She put her sentence down and everyone read along as she read out loud.

Student 2: "So my sentence is the second one then, because it talks about how they spend their money."

Student 4: "My sentence talks about how they spend their money too."

Student 2: [sigh] "Okay. So, you go next. I guess I have the third sentence."

Here I guided thinking once again. "So, both sentences talk about how money is spent. Are there any differences?"

Student 4: "Wait... my sentence doesn't talk about a money amount. It says 'the rest of their money.' Does your sentence talk about money?"

Student 2: "Yeah."

Student 1: [Points to Student 2] "Then you are the second sentence, not him." Student 2 read his sentence aloud and placed it beside sentence one for the group to see.

Student 4: "All right, this is easy. I have the third sentence because it is the amount that we have to write the equation for." He read his sentence aloud and the group rearranged the sentences so that they were vertical instead of horizontal.

I had struck gold with this group. The students were collectively responsive with their reading, attending to the text, their own thoughts, and the thoughts of their group (Beers and Probst 2017). Student 3's question, "Who is *they*?" influenced the depth of the focus on the text and the group's collective thought process. It was because of their responsiveness that Student 4 was able to comment that his portion of the task was easy. The collection of ideas enriched the meaning of the text and yielded the coveted high-quality experience with text that is so critical to effective reading instruction (Fountas and Pinnell 2006). This second group possessed the mathematical foundations and conceptual knowledge and collectively addressed the structure of the word problem using components of reading instruction. Consequently, they successfully wrote an equation for *s*.

Even as a mathematics teacher, components of reading instruction must be attended to when comprehending word problems. Your job becomes two-fold: nurture mathematical foundations and conceptual understandings *and* foster components of reading instruction that must be attended to when comprehending word problems.

REVISED JIGSAW WORD PROBLEM PROCEDURE

After shifting perspectives, and with conceptual orientation as a focus, this is how I restructured the jigsaw word problems procedure:

1. Rewrite the word problem so that each sentence is isolated on its own sentence strip.
2. Students sit in groups of three (if the problem is three sentences long) or four (if the problem is four sentences long).
3. Place the sentence strips face down in the middle of the group and let each student choose a sentence strip.
4. Students do not share with their group but may ask for

assistance as needed from an outside party while reading their sentence. Be sure sentence stems are accessible to provide academic language support.

5. After everyone in the group has read their sentence to themselves, they begin piecing together their problem. The person who believes he/she has the last sentence of the problem says, "I think I have the last sentence of the problem because..."

Note: When students justify their answers, they do *not* read their sentence aloud. This makes them actively justify their reasoning instead of passively reading.

   a. Example: *I think I have the last sentence because it asks for an equation to solve a missing amount.*
   b. Non-example: *I think I have the last sentence because it says, "What is an equation that can be used to find* s, *the amount of money in dollars the friends spent on snacks?"*

6. If everyone in the group agrees that the student does have the last sentence based on his/her reasoning, the student then reads the actual sentence aloud to the group and places it in the middle of the table.
   a. If someone else in the group believes he/she has the last sentence, that person then says, "I think I have the last sentence of the problem because..."
   b. The group, after listening to the two explanations, decides which person has the last sentence. That person then reads the actual sentence aloud and places it in the middle of the group for all to see.
7. The person who believes he/she has the first sentence then says, "I think I have the first sentence because..."
   a. Example: *I think I have the first sentence because it states a total, and we know we will be looking for a part.*
   b. Non-example: *I think I have the first sentence because it says, "Five friends spent $105 at the movies."*

c. Follow the same procedure for step 6. After reading the actual sentence aloud, the student will place the sentence to the right of the first sentence.

8. The person who believes he/she has the second sentence then says, “I think I have the second sentence because...”
    a. Example: *I think I have the second sentence because the first sentence mentions how much money they have, and my sentence mentions how much money they spent. (Now I know “they” is friends, and they are at the movies.)*
    b. Non-example: *I think I have the second sentence because it says they spent $50 on tickets and $21.25 on drinks.*
    c. Follow the same procedure for step 6. After reading the actual sentence aloud, the student will place the sentence to the right of the first sentence.
9. The student who thinks he/she has the third sentence says, “I think I have the third sentence because...”
    a. Example: *I think I have the third sentence because the first two sentences mention a whole amount and partially spent amounts. My sentence mentions how the rest of the money is spent. It is another partial amount.*
    b. Non-example: *I think I have the third sentence because they spent the rest of their money on snacks.*
    c. Follow the same procedure for step 6. After reading the actual sentence aloud, the student will place the sentence to the right of the second sentence.
10. The students then engage in academic discourse about how to solve the problem. They record their answer.

You might have read this whole jigsaw activity and do not see a value in it for mathematics instruction. You might be wondering how to get this level of responsiveness from a group of students who have not had exposure to this method of decomposing text. The value of this strategic approach focuses on the three skills students need to comprehend word problems:

- understand the relationship between quantities,
- embrace academic conversations to express and gain problem-solving perspectives, and
- differentiate between answering and making sense of the mathematics.

A word problem requires reading—not skimming for information—to comprehend the relationship between numbers. A single sentence can expose relationships or possible relationships between quantities. For example, "The rest of their money was spent on snacks" lets the mathematical reader know that there is a quantity for the total money amount and the snacks are a smaller amount taken from this total. There is a value to taking the time during mathematics instruction to foster reading comprehension skills to support quantitative analysis.

Reflection Checkpoint
What possible quantities can be explored with the sentence, "Five friends spent $105 at the movies"? How can you guide students to discover these possible quantities?

I want you to have success in your classroom. Appendix D has student instructions to the jigsaw task and two word problems you can use with students that are mathematically below their threshold. It is the structure of the task and academic discourse that will be the challenge. (If the problems are still too difficult mathematically, please change the problems to meet your students' needs.) The purpose is to give students the opportunity and practice to familiarize themselves with the structure. When they no longer have to think about the structure of the task, they can shift focus to solving challenging problems collectively.

If you are unsure of how to begin or have failed in your first or second attempt as I had, I encourage you to reach out to your coworkers or other educators. Talk about the experience and learn from one another as you refine your instruction and planning to meet the needs of your students. This approach to word problem-solving demands a shift in thinking and communicating. Change takes time. It is the learning throughout the change that makes the shift worthwhile.

One area of learning might be that students need to be aware of and name their thinking for purposeful conversation and social learning to occur (McGregor 2007). Students with limited exposure to verbalizing metacognition might need additional language support. This may take a lot of practice. Students should be able to reference exemplar language as they learn to name their thinking:

- This sentence introduces the whole.
- This sentence is asking for a whole.
- This sentence mentions equal groups.
- This sentence is asking for equal groups.
- This sentence has unequal amounts.
- This sentence mentions parts of a whole.
- This sentence is asking for a part of a whole.
- This sentence is asking for an unknown remaining amount.
- This sentence is asking for a total amount.
- This sentence is not asking us to solve an equation.
- This sentence is asking us to write an equation.

You and your students will be able to write customized sentences that name thinking to meet the needs of your classroom. The above sentences are just examples from which to start that I used in classrooms. *Solving Word Problems: Developing Quantitative Reasoning* (Charles 2011) provides an extended list of resources for greater academic language support.

In my two student group examples, the first group defaulted to adding all the numbers together to solve the word problem. They had failed to analyze the relationships between numbers, did not respond to the text or one another, and therefore did not apply metacognition. Because of their lack of comprehension of the words and consequently the relationships that support quantitative reasoning, they solved for an imaginary equation when what the problem required was to *write* an equation. In contrast, the second group began with an understanding that they were writing an equation. They worked off one another's responses, questioned, and collaborated to comprehend the words and performed a quantitative analysis and reasoning. They successfully wrote an equation.

FURTHER APPLICATION

The above scenario happened about three years prior to my research for this book. I have always enjoyed applying ideas and discovering why they work or don't work, and I was fortunate to have recorded this. However, while researching, I decided to try the jigsaw word problem again. This time I used a much younger group, with mixed ages, reading abilities, and conceptual mathematics knowledge. I was curious if this jigsaw word problem task could still yield academic conversations of engaged, varying perspectives in a group with an even larger discrepancy of foundational understanding. The three students were a ten-year-old (Student A), eight-year-old (Student B), and seven-year-old (Student C). Student A, a fourth-grader, is diagnosed dyslexic and struggles with reading fluency but not comprehension. Student B, a third-grader, does not have strong number sense and struggles with reading fluency and comprehension. Student C, a first-grader, is the youngest but excels in mathematics and reading comprehension. The three students are cousins and comfortable with one another. Here is their word problem:

Michelle practiced swimming the 100-meter race three times yesterday. Her times were 18.175 seconds, 18.33 seconds, and 18.4 seconds. What was Michelle's total time in seconds for the three practice races?

This is what transpired:

Me: What is a word problem?

Student B: It's words for a problem.

Student A: There's keywords in a problem to let you know what it is.

Student C chose to remain silent as his cousins answered my question. As a first-grader, there is a possibility he was unfamiliar with the term *word problem*.

Next, I explained the process to the jigsaw word problem activity, but this time, I asked them to write what they wanted to say before they spoke. I specifically told them that they would not have to solve the problem, only put it in order. Below is the problem and what each student wrote. The sticky notes are from moments when they asked me how to spell a word and I jotted it down.

Student A: I thing it is the end becaue it asking you do to so thing. [You will see her attempt to solve the problem, but this occurred after the task finished.]

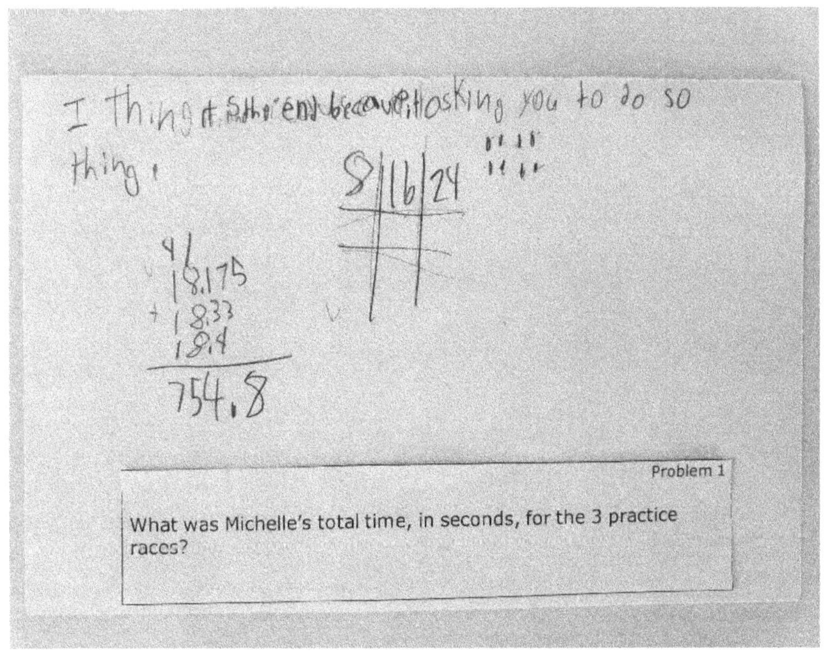

Student B: I think I have the ending because my sentence has a period at the end of the sentence.

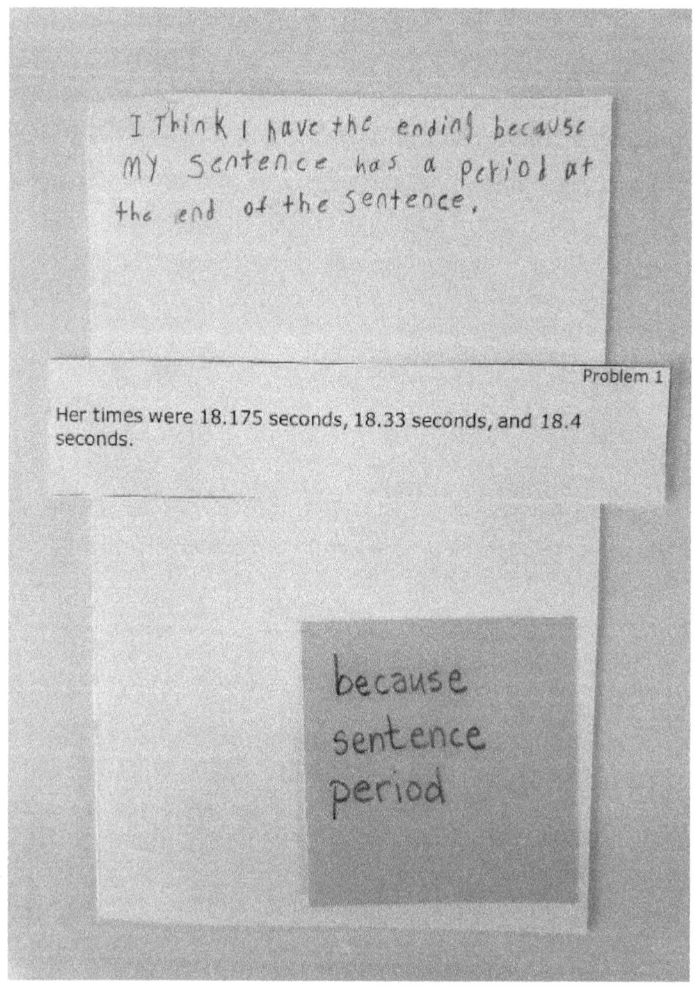

Student C: i think its the second beacause it says michelle practiced swimming yesterday.

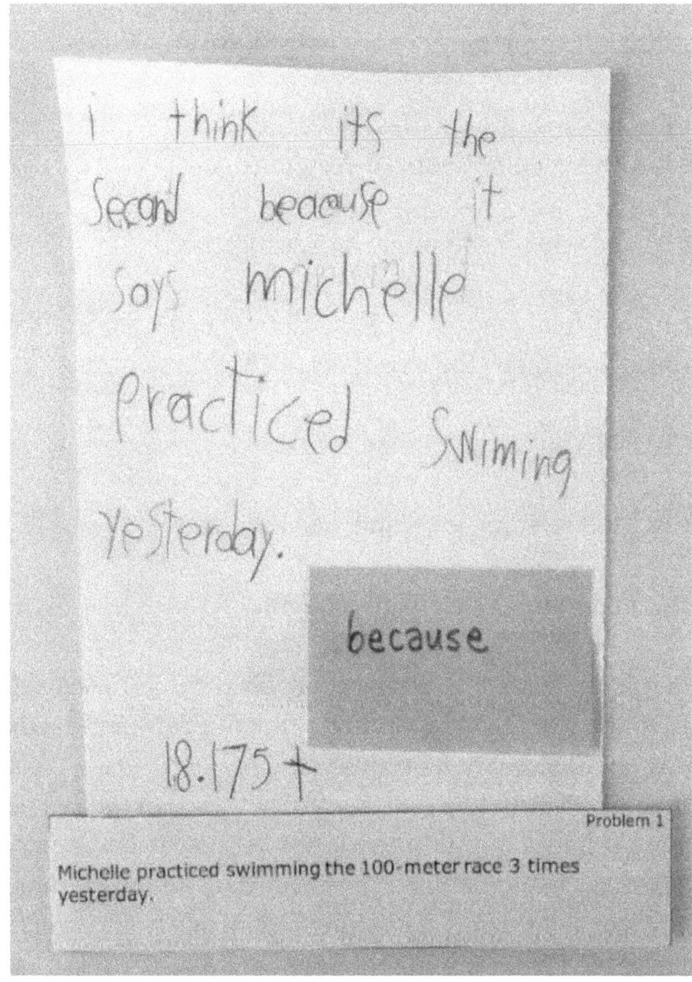

I was surprised that the students were able to justify themselves without restating their sentence. When I worked with the fifth graders, this was an area in which we had to practice multiple times. When they began discussion, they read their notes aloud. After they each read aloud, this conversation happened without my prompting.

Student A: In other parts of the problem they are telling things, but this is asking for something. That's why I have the ending.

Student C: Then that means that I have the second sentence because... because it says Michelle practiced yesterday.

Student B: And when I add mine, it goes together.

Student A: But his [Student C] has the first name, so in a book the name comes first, then [the word] *her*.

Student C: Oh! You're right, mine *is* first.

Student B: Okay so mine is second, like I thought.

Student A: And mine is last, like I thought.

Notice in this conversation, Student A applied her knowledge of story structure to the word problem genre. Student C, also interacting with text, acknowledged Student A's insight and grew from her perspective. Student B, however, wrote that she had the last sentence not because it had a question mark but because it had a period. No one addressed this. She later claimed that she had the second sentence, "like I thought." This interaction made me wonder how much she learned from the conversation.

The three students put their sentences in order and reread the whole problem. I asked them what they needed to do to solve the problem.

Student C: I think add.

Student A: I think it's add because it is asking for the total.

Student C: Yeah, it's asking for the total.

Me: What are you adding?

Students A and C in unison: 18.175 + 18.33 + 18.4.

Me: How do you know you add these numbers?

Student C: Because it says there are three [practice races].

Notice that Student B, the one who struggles with number sense, stayed quiet during the discussion of the process for solving. I observed her during this time. Her eyes would move from the speaker to the word problem. While the other two were working slightly below their threshold of understanding–that is, they already understood they would need to add and why–Student B did not have this understanding. Although silent, she was exposed to language and acquisition of concept, and her eyes at least were actively engaged.

I think it is important to note that Student A, who has been taught to use keywords, mentioned the word *total* for the reason to add. In this case, her statement is reasonable. I cannot help but wonder, had she gotten Student B's sentence: *Michelle practiced swimming the 100-meter race three times yesterday*, if she would have argued that they would need to multiply because of the word *times*.

Something else I must point out is that even though Students B and C had yet to be introduced to decimals based on state standards, neither of them mentioned not understanding them during the jigsaw task. Student B might not have mentioned the decimals because she was experiencing cognitive overload in which she was overwhelmed by how much information she needed to process in a short time. Student C, however, understood the numbers to be values assigned to practice times and only inquired how to add them when he wanted, on his own accord, to solve the problem.

It would be interesting to continue working with these three cousins and monitor the change in Student B as she is exposed to more academic discourse in this task. However, my purpose of the activity was to see if a group of students with greater discrepancies—such as age and grade levels, exposure to concepts, reading abilities, and numeracy—could still create their own zone of proximal development. The answer is a cautious yes because I had eliminated a significant variable. Had the students not already felt like they were in a safe learning environment, I wonder if they would have been as willing to express their thoughts or as quick to learn from one another.

I mentioned in Chapter 1 that conceptual orientation may take a full content block to attain, but that it is worth it. This comment comes from experience. I must forewarn, it took me an entire mathematics block to successfully practice the structure for the jigsaw activity. The time it took to introduce the task was a long-term investment. Urie Bronfenbrenner's (1979) social-ecological theory of human development as cited by Walqui and van Lier (2010) states, "ideas emerge in interaction, are shared with peers, and are further developed in the interaction" (29). Structured conversations, while awkward at first, foster the academic language students need to explain perspectives and insight, deepening conceptual understanding through social learning. Students will move from underlining and circling keywords and questions to reading a problem, performing a quantitative analysis and reasoning, understanding their thought process, verbalizing and sometimes defending their metacognition, absorbing the perspectives and voices of others, and fully comprehending what it is they must accomplish. These are all skills applicable beyond a state-mandated test.

Concepts Summary

- Students must become responsive readers to comprehend text.
- A single sentence can reveal relationships between quantities. Students might need language support to express these relationships.
- Through purposeful interaction, students can express ideas and build on them to accomplish a goal.

CHAPTER 5

CREATING A CLASS CULTURE

Each of the three approaches—taking away the mathematics to understand the mathematics, performing a quantitative analysis to build a problem, and jigsaw word problems—revolve around a framework of social learning within purposeful instruction. They all include reflective thinking and justification of thought as well. These skills within small group or whole group discussion require acquisition of academic language. For all of this, creating a community of learners is ideal.

In her book *Culturally Responsive Teaching and the Brain,* Zaretta Hammond states a toxic, hostile, or unwelcoming environment prevents learners from producing enough oxytocin, resulting in anxiety and triggering the nervous system because "the brain doesn't experience a sense of community" (2015, 45). Without this sense of a safe environment, learning is not possible; the need to build positive social relationships within a diverse classroom culture is related to the learning that takes place. To build these relationships, Hammond identified culturally responsive brain principles that work together non-linearly, except for the first principal, which must be established for the others to occur (2015). I will paraphrase her principles and encourage you to read her work for a deeper understanding.

1. The brain desires connection and will seek ways to diminish social threats while expanding opportunities for this connection.
2. The brain and nervous system will remain in check through

positive relationships.

3. How we process information has a cultural foundation.
4. Learning requires deliberate processing, stemmed from active and focused attention.
5. The brain builds on prior knowledge to attain new knowledge.
6. Challenges expand the brain's ability to think with complexity.

Established socialization in a safe environment provides students with the opportunity to practice academic discourse and take risks with newly acquired vocabulary, phrases, or thoughts. Because the learner necessitates both an opportunity and a safe environment to verbalize thought processes, it is imperative for the teacher to create a classroom culture to provide and promote these two conditions.

Cummins argues that interactions between students and teachers are more crucial to student achievement than any method for teaching content (2001). As the leader of your classroom, you set this expectation of mutual respect through interaction (Parrish 2014). It is in the classroom that teachers can send micro-affirmations to promote the positive student self-image as thinkers, learners, community contributors and self-efficient communicators. One starting point is to build trust to cement a sense of connection and student belief that what he or she has to say is valued.

Hammond details trust generators in her book (2015), and I encourage you to read it for a greater understanding of what I will again paraphrase. She describes five ways to build trust through *selective vulnerability*, or sharing your struggle; *familiarity*, or meeting often; *similarity of interests*, or sharing interests; *concern*, or asking follow-up questions about someone's life; and *competence*, or demonstrating the ability to teach (Hammond 2015, 79). From a pedagogical viewpoint, cultivating trust helps the teacher relate to his or her students better and fosters a deeper purpose for teaching as it builds on personal investment. Once a connection of trust is established, these norms become embedded in the classroom culture and students are safe to fail, learn, and succeed with their peers and teacher.

Reflection Checkpoint
What are micro-affirmations you can use with students intimidated by mathematics?

In student collaboration in a safe environment, look for:

- Student-led academic conversations
- Comfort in collaboration
- Comfort in being wrong

During student collaboration in a safe environment, look for a teacher who:

- Uses active listening to student ideas
- Models language and thinking
- Acknowledges the struggle
- Provides micro-affirmations

For a safe classroom environment, look for:

- Evidence of every student voice posted in the classroom
- Isolated areas are for students to go to by choice
- Evidence of positive micro-messaging; students see themselves and their identities positively in classroom exemplars, posters, books, etc.

Another component to classroom culture is the understanding that everyone can learn from each other. Have you ever been in a professional environment in which you did not express what you were thinking? What kept you from sharing your perspective with the group? Did you not believe that what you had to say was a value from which others could learn? Did you not believe others would listen?

Understanding and respecting that everyone learns from one another is especially important to mathematicians because they share, test, and develop their ideas collaboratively (Ferrini-Mundy 1996). If students understand that they can learn from one another, then they also understand that how others think is not always the same. Visualizing different thought processes emphasizes the various approaches a learner can take while problem-solving. However, acquiring the skills for metacognition, the academic language and process standards do not come in a one-size-fits-all mold. In her book *Number Talks: Whole Number Computation,* Parrish suggests six ways to develop student accountability while honoring the individual's learning path (2014). All six suggestions are noteworthy, but for the purpose of *this* book, I will mention just two that revolve around building a classroom culture of reflective learners. She suggests 1) keeping a record of presented problems and the strategies students employed to solve them as well as 2) posting class strategy charts (Parish 2014). Both suggestions respect and value the learners' approaches to problem-solving and allow students to reference their previous visual thought processes for similar word problems.

Years ago, I visited a classroom where the teacher posted a laminated chart paper with all her students blocked on it by sticky notes. She would present a word problem on Monday that the students needed to solve by Friday. The students would write in dry erase markers how they solved the problem in their sticky note block on the chart paper. Giving the students a week to solve the problem was an innovative approach to perpetuating a safe learning environment. Students who did not know how to solve the problem were able to learn from other students' written submissions during the week before posting their own response. Here is a visual mock-up of this teacher's chart:

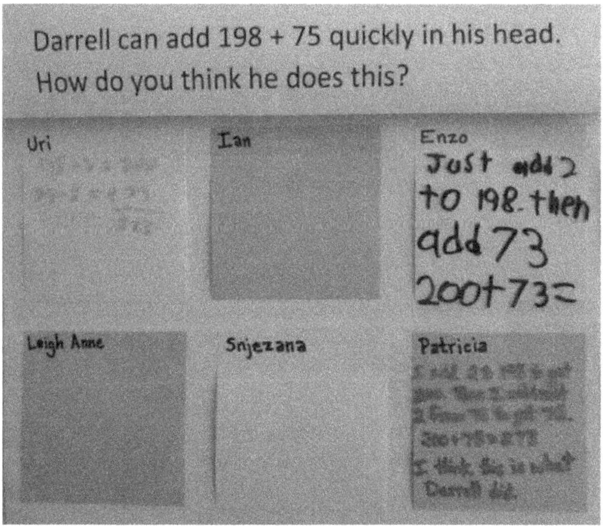

In student collaboration with respect to perspective, look for:

- Respectful questioning of one another
- Respectful clarification of ideas/concepts
- Respectful challenges to ideas/concepts
- Respectful acknowledgment of ideas and building on those ideas

During student collaboration with respect to perspective, look for a teacher who:

- Uses active listening to student ideas
- Asks questions that promote learning and conceptualization instead of telling
- Clarifies/addresses misconceptions

For student collaboration with respect to perspective in the classroom environment, look for:

- Visual representation of student voices that build on one another
- Seating is arranged so students can work with one another face-to-face
- The environment is full of purposeful noise

A safe environment in which to learn and an understanding that different perspectives mold and strengthen understanding create a foundation for inclusion and invitation for discovery. From this foundation, the culture of the classroom will embrace diverse approaches to problem-solving from every student and perpetuate positive self-images of abilities and identities as capable thinkers. This respect for and valuing of every voice is a component to culturally responsive teaching. As exemplified with Students A, B, and C in the jigsaw approach, differing foundations in literacy and mathematics and even grade levels do not prohibit students from learning from one another because they respected and valued each other's voices in a safe learning environment.

Concepts Summary

- Teachers are the leaders in the classroom and set environmental expectations.
- Empower student voice through student discourse, anchor charts, verbal acknowledgments, and displayed student work, etc.
- Students with varying degrees of literacy and mathematical foundations will have a more positive self-image when placed in a safe environment.

FINAL THOUGHTS

While writing this book was still in its infancy, I pitched my ideas and research to my youngest brother during a five-hour road trip. He listened attentively and asked probing questions, but at the end of our discussion, he made a comment that I have heard so many people say.

"That's great and all, Sister, but word problems just don't happen in everyday life."

It is important to note that there is a difference between generating questions that take away the mathematics and acknowledging actual word problems in real life. However, for a guy that creates codes in spreadsheets to manipulate data *in his spare time*, I was surprised that he did not see the application and occurrence of word problems all around him. I think it may be our disassociation of words and calculations that prevent us from seeing connections, but that is only a surface thought. He had given me a challenge. Our conversation happened in mid-October. Two weeks later, my brother shared his homemade Halloween costume in our sibling group chat.

He was Zoltar, the fortune teller from the movie *Big* that dispenses cards with your fortune. Here was our conversation thread following his pic:

Me: Did you tell many fortunes?

Brother: I passed out all but about twenty cards.

Me: How many did you print?

Brother: I made three pages with four columns and fourteen rows each.

Me: This is a word problem.

Brother: Damn you.

Not long after this, we were at my parents' house eating breakfast. My youngest son was eating breakfast sausage links with rice. I asked him how many he had eaten and before he could answer, my mother replied, "Well, I gave him five links, but I took two."

I smiled and turned to my brother. When he looked over at me, I said, "That seems to be another word problem that happened in everyday life." He let out a defeated laugh, and I jotted down what had just transpired so I could add it to my examples.

Like with my brothers, my sons and I like to use quantitative reasoning to communicate. Recently my seventeen-year-old asked me if he should bake some bread for me to eat with soup. I told him not to go through the trouble since we have crackers. The morning after that conversation I woke up to this message on our family whiteboard:

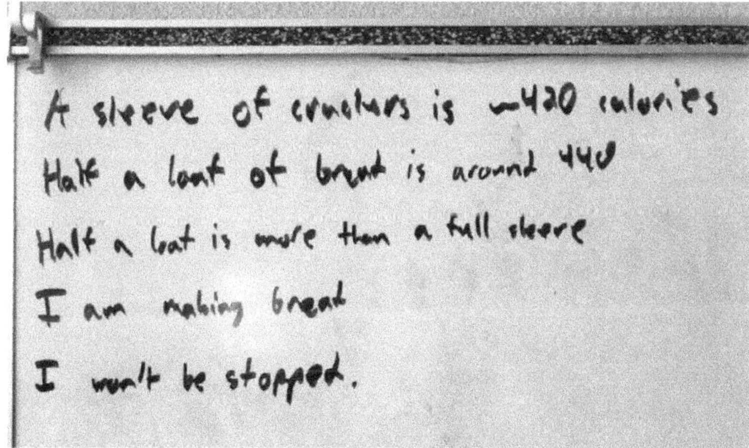

If you can't read his handwriting, it states, "A sleeve of crackers is ~420 calories. Half a loaf of bread is around 440. Half a loaf is more than a full sleeve. I am making bread. I won't be stopped."

Here are other examples that I have come across from conversations over the years:

- "How old will Mom be on her birthday?"
  "I can't remember, but we know she was born in 1960."
- "We got here at 11:45. How long have we been waiting for our table?"
- "Maya and Ariana live about 300 miles away. How much gas money do we need?"
- "How many days do I have to work to buy my headphones?"
- "I wonder what percentage of my day is spent taking care of—"
- "This shirt is 15% off and these shorts are 30% off. Do I have enough money to purchase both?"

I hope that as you read these examples, a few popped up in your head as well. And, because of the frequency illusion, I hope that you will notice many more in the future. This mindset that word problems are not organic is false. We just need to be mindful of their presence.

APPENDIX A

Λυισ βουγητ 2 τιχκετσ το α χονχερτ. Ατ τηε τιμε οφ πυρχηασε, νινε ηυνδρεδ τηιρτψ–σεϖεν σεατσ ρεμαινεδ ιν αν αυδιτοριυμ οφ 10,000 σεατσ. Εξπρεσσ τηε αττενδανχε ατ α περχεντ οφ φιλλεδ χαπαχιτψ ατ τηε τιμε οφ Λυισ πυρχηασε.

Luis bought 2 tickets to a concert. At the time of purchase, nine hundred thirty-seven seats remained in an auditorium of 10,000 seats. Express the attendance at a percent of filled capacity at the time of Luis's purchase.

APPENDIX B

Quantitative Reasoning Prompts

Identifying a Quantity	Identifying a Value
Does my sentence mention something that can be measured or counted?	What is the measurement or the number value of —?
Does my sentence mention more than one thing that can be measured or counted?	Is there a missing value for a quantity?
Is there something I need to find that needs to be measured or counted?	Are there quantities related to other quantities that can help me find an unknown value?

Example Response Stems

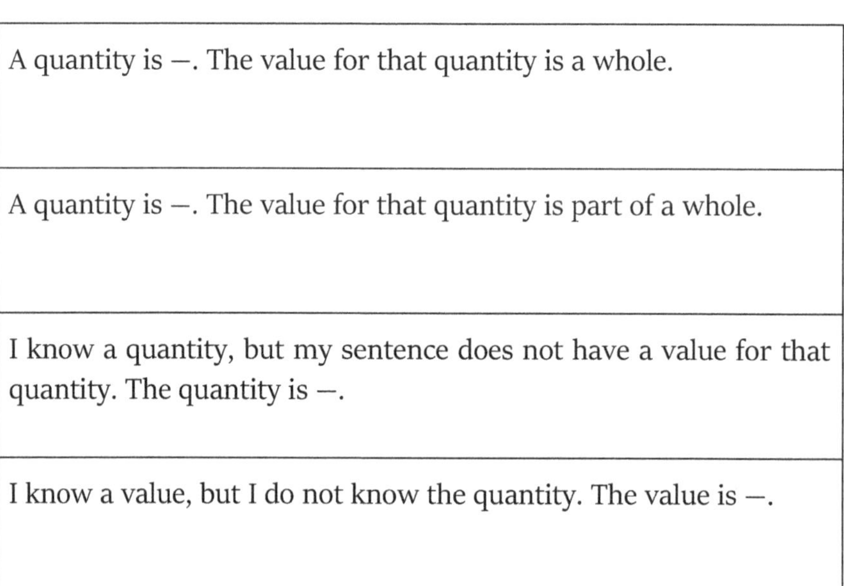

A quantity is —. The value for that quantity is a whole.
A quantity is —. The value for that quantity is part of a whole.
I know a quantity, but my sentence does not have a value for that quantity. The quantity is —.
I know a value, but I do not know the quantity. The value is —.

The word (e.g., *each* or *same number*) implies two or more equal parts. I have the value for one equal part of the quantity —.

Example response stems for the final sentence

My sentence is asking to compare two quantities. The value of this comparison is unknown.
My sentence is asking to join two quantities. The value of the joined quantities is unknown.
My sentence is asking for the equal value of a quantity. The other value and total are known.
My sentence is asking for a total of equal values.
My sentence is asking for an equation to find the value of the quantity —.

APPENDIX C

Example problems with a missing last sentence

Charlotte and Olivia are playing a board game. Charlotte has $1,020. She must pay Olivia $510.
Tad made 5 dozen mochi because he was bored. He gave 2 dozen mochi to his cousin Tee. He gave half as many mochi that he gave Tee to his other cousin Mai.

APPENDIX D

JIGSAW WORD PROBLEM INSTRUCTIONS FOR STUDENTS

The purpose of this task is to piece together a word problem and then solve it as a group.

1. Mix up the sentence strips and place them face down in the middle of your group. Each sentence strip is one sentence in your word problem.
2. Choose one sentence strip each.
3. Read your sentence, but do not share your sentence with any other group member! You can draw if it helps you. You can ask for help with your sentence if you need it.
4. After every group member has read their sentence to themselves, your group will begin piecing together your word problem. You can use sentence stems to help you explain.
5. It is the choice of the group to either start with the first sentence or start with the last sentence. (Because it often ends in a question mark, the last sentence is easy to identify.) If you start with the last sentence:
   a. The person who believes he or she has the last sentences states, "I think I have the last sentence because it ends in a question mark. It wants us to solve for..."
   b. If all the students agree that the person talking does have the last sentence, then that person can finally put the sentence down in the middle of the group and read it aloud as the group reads along.
   c. NOTE: *You may not read your whole sentence aloud to the group until everyone agrees that your sentence is the correct one. You can use sentence stems to help you talk about your sentence.*
6. The person who believes he or she has the first sentence says, "I think I have the first sentence because..."
   a. Example: "I think I have the first sentence because it talks about a total number."
7. The person who believes he or she has the second sentence

says, “I think I have the second sentence because...”

   a. Example: “I think I have the second sentence because it talks about an equal part of the total or whole.”

8. The person who believes he or she has the third sentence says, “I think I have the third sentence because...”
   a. Example: “I think I have the third sentence because the first two sentences talk about a whole and an equal part of a whole. My sentence talks how many equal parts she needs.”
9. When all sentences are laid out, your group can take another look. You can talk about rearranging a sentence if it doesn’t make sense. (Ask your teacher for help if it gets confusing.) If your group likes the order, talk about the problem and solve together. Feel free to draw and work on scratch paper.

JIGSAW WORD PROBLEMS

Problem 1
Sonora earned $108 babysitting her cousin.

Problem 1
She spent $49 on food.

Problem 1
She put the rest of her money in her savings account.

Problem 1
How much money did Sonora save?

Problem 2 Kai goes to basketball practice five times a week.
Problem 2 Each practice is two hours long.
Problem 2 Last week he missed one practice.
Problem 2 How many hours did Kai practice basketball last week?

REFERENCES

Beers, K., and R. E. Probst. 2017. *Disrupting Thinking: Why How We Read Matters*. New York: Scholastic.

Charles, R. 2011. *Solving Word Problems: Developing Quantitative Reasoning*. London: Pearson.

Clement, L., and J. Bernhard. 2005. "A Problem-Solving Alternative to Using Key Words," *Mathematics Teaching in the Middle School* 10(7).

Cummins, J. 2001. *Negotiating Identities: Education for Empowerment in a Diverse Society*. Los Angeles: California Association for Bilingual Education.

Driscoll, M., J. Nikula, and J. DePiper. 2006. *Mathematics Thinking and Communication: Access for English Learners*. Portsmouth, NH: Heinemann.

Ferrini-Mundy, J. 1996. 'Mathematical Thought-in-Action: Rich Rewards and Challenging Dilemmas' in *What's Happening in Math Class?: Envisioning New Practices Through Teacher Narratives*. New York: Teachers College Press.

Fountas, I., and G. Pinnell. 2006. *Teaching for Comprehension and Fluency: Thinking, Talking, and Writing about Reading*. Portsmouth, NH: Heinemann.

Fry, Edward. 2000. *1000 Instant Words*. Garden Grove, CA: Teacher Created Resources, Inc.

Hammond, Z. 2015. *Culturally Responsive Teaching & the Brain: Promoting Authentic Engagement and Rigor Among Culturally and Linguistically Diverse Students*. Thousand Oaks, CA: Corwin.

Hernandez, A. 2020. Personal communication, March 4, 2020.

Hernandez, I. 2020. Personal communication, March 4, 2020.

Kue, H. 2019. Personal communication, October 19, 2019; November 2, 2019; November 9, 2019.

Land, R., J. H. F. Myer, and M. Flanagan. 2016. *Threshold Concepts in Practice*. Leiden, The Netherlands: Brill.

Martinez, G. 2019. Personal communication, October 30, 2019.

McGregor, T. 2007. *Comprehension Connections: Bridges to Strategic Reading*. Portsmouth, NH: Heinemann.

Meyer, Dan. 2010. "Mathematics Class Needs a Makeover," *TED*.

YouTube, March 6, 2010. https://www.ted.com/talks/-dan meyer mathematics class needs a makeover/transcript?language=en.

Parrish, S. 2014. *Number Talks: Whole Number Computation.* Sausalito, CA: Mathematics Solutions.

Simon, M. 1996. 'Focusing on Learning Mathematics' in *What's Happening in Math Class?: Envisioning New Practices Through Teacher Narratives.* New York: Teachers College Press.

State of Texas Assessments of Academic Readiness (STAAR®). 2017. Austin, TX: Texas Education Agency.

Thompson, A. G., R. A. Philipp, P. W. Thompson, and B. A. Boyd. 1994. "Calculational and Conceptual Orientations in Teaching Mathematics." In *1994 Yearbook of the NCTM*, edited by A. Coxford, 79–92. Reston, VA: NCTM.

Vaughn, S., and S. Linan-Thompson. 2004. *Research-Based Methods of Reading Instruction: Grades K-3.* Alexandria, VA: Association for Supervision and Curriculum Development.

Walqui, A., and L. van Lier. 2010. *Scaffolding the Academic Success of Adolescent English Language Learners: A Pedagogy of Promise.* San Francisco, CA: WestEd.

Webb, Noreen M., Megan L. Franke, Marsha Ing, Jacqueline Wong, Cecilia H. Fernandez, Nami Shin, and Angela C. Turrou. 2014. "Engaging with others' mathematical ideas: Interrelationships among student participation, teachers' instructional practices, and learning." *International Journal of Educational Research* 63: 79–93.

ACKNOWLEDGEMENTS

Diane would like to thank Leigh Anne and Kyle for their guidance and support while writing. Leigh Anne was the first person to read her manuscript and offered perspective. Kyle offered guidance and direction. How much thanks should she give?

Quantitative Analysis:

The amount of thanks to Leigh Anne	Immeasurable value of thanks
The amount of thanks to Kyle	Immeasurable value of thanks

ABOUT ATMOSPHERE PRESS

Atmosphere Press is an independent, full-service publisher for excellent books in all genres and for all audiences. Learn more about what we do at atmospherepress.com.

We encourage you to check out some of Atmosphere's latest releases, which are available at Amazon.com and via order from your local bookstore:

Convergence: The Interconnection of Extraordinary Experiences, by Barbara Mango, Ph.D., and Lynn Miller, MS

Sacred Fool, by Nathan Dean Talamantez

My Place in the Spiral, a memoir by Rebecca Beardsall

My Eight Dads, a memoir by Mark Kirby

Dinner's Ready! Recipes for Working Moms, by Rebecca Cailor

Vespers' Lament: Essays of Culture Critique, Future Suffering, and Christian Salvation, by Brian Howard Luce

Without Her, a memoir by Patsy Creedy

One Warrior to Another, a reflection by Richard Cleaves

Emotional Liberation: Life Beyond Triggers and Trauma, a guide by GuruMeher Khalsa

The Space Between Seconds, by NY Haynes

License to Learn: Elevating Discomfort in Service of Lifelong Learning, by Anna Switzer

The Bond, a memoir by A. M. Grotticelli

Sex—Interrupted: Igniting Intimacy While Living With Illness or Disability, by Iris Zink, BSN, and Jenny Palter

Between Each Step: A Married Couple's Thru Hike On New Zealand's Te Araroa, by Patrice La Vigne

Waking Up Marriage: Finding Truth Inside Your Partnership, by Bill O'Herron

ABOUT THE AUTHOR

Diane Kue has known she wanted to be a teacher since first grade. Growing up, her perspective as a student was unique. She learned so she *could* teach. She became a teacher in 2003 and taught for ten years before transitioning to curriculum writer and instructional support for teachers of emergent English Learners. She created and coordinated special programs for student success. She has presented locally at technology conferences and principal summits; regionally and state-wide at the TCU Racial Equity Summit and the Conference on the Education of Hispanics; and nationally for the National Council of Teachers of Mathematics.

CPSIA information can be obtained
at www.ICGtesting.com
Printed in the USA
BVHW042300211022
649995BV00021B/217

9 781637 529027